CONTROLLING
TECHNOLOGY

CONTROLLING TECHNOLOGY

Genetic Engineering and the Law

Yvonne M. Cripps

PRAEGER

PRAEGER SPECIAL STUDIES • PRAEGER SCIENTIFIC

Library of Congress Cataloging in Publication Data

Cripps, Yvonne M
    Controlling technology.

    Includes bibliographical references and index.
    1. Recombinant DNA--Research--Law and legislation.
I. Title.
K3611.G46C75   344'.095   80-13754
ISBN 0-03-056806-4

Published in 1980 by Praeger Publishers
CBS Educational and Professional Publishing
A Division of CBS, Inc.
521 Fifth Avenue, New York, New York 10017 U.S.A.

0123456789   038   987654321

Printed in the United States of America

To Una and Robert

ACKNOWLEDGMENTS

Many individuals and organizations participated in the preparation of this book and I am greatly indebted to all of them. In particular, very special thanks are due to Professor K. J. Keith, Dean of the Faculty of Law at Victoria University, Wellington, New Zealand. His guidance and comments were of immense value in enabling me to write the thesis on which this text is based. I am also very pleased to record my appreciation of the encouragement I received during the latter stages of this project from Mrs. Dianne Sleek of the Faculty of Law at Victoria University.

CONTENTS

INTRODUCTION

Recent discoveries in the field of molecular biology that is popularly known as "genetic engineering" have given rise to a considerable amount of concern and debate both among scientists and in the wider community. There have been persistent calls for the control of this new technology and the technologists who manipulate the genetic material of plant and animal life.

This book will focus on recombinant DNA and somatic cell genetic techniques. The procedures involved represent closely related facets of a wider ability to "engineer" genetic composition. In a broad sense, that skill includes the older and more familiar techniques of selective breeding and genetic screening. Recombinant DNA and somatic cell genetic studies are, however, of special importance because they pose unique and unpredictable physical threats to human life, to the environment, and to countries with agriculturally based economies. Ironically, those threats may be counterbalanced by important medical, environmental, and agricultural benefits.

An accident that involved a genetic engineering technique has already occurred in New Zealand. It arose out of a protoplast fusion experiment* that was designed to improve the nitrogen-fixing capacity of a fungus that is commonly associated with pine trees. The modified fungus was introduced to *pinus radiata* seedlings at the Plant Physiology Laboratory of the Department of Scientific and Industrial Research (DSIR) in Palmerston North, and within a few weeks all the seedlings that were associated with the modified fungus had died. Biological containment measures that would have helped to minimize the possibility that this fungus could exist outside the laboratory were not implemented. Fortunately, no field trials were undertaken and the fungus does not appear to have escaped from the laboratory.

From the legal viewpoint the lesson that is most forcefully illustrated by the incident is that the researchers involved had complete discretion to devise and conduct an experiment that had the potential to destroy or

---

*Protoplast fusion is one of the somatic cell genetic techniques. See Chapter 1 and the Glossary.

severely damage a primary industry. They exercised that discretion unfettered by formal legal controls and in the absence of procedural guidelines.[1]

This book is devoted to an analysis of legal controls on genetic engineering both in the public and the private sectors. In the public sector the DSIR and several New Zealand universities are engaged in genetic engineering research.* The Minister of Science and Technology has requested those institutions, and various other organizations — including the Dairy Research Institute, the Departments of Health, Agriculture and Fisheries, and the New Zealand Forest Service — to observe combined DSIR/Medical Research Council (MRC) guidelines.[2]

The difficulty with these "requests" is that the guidelines are nonmandatory. The problem is heightened in the private sector where companies are manufacturing and using genetically engineered organisms and applying for patents on new products and processes.[3] Those organizations do not rely on government funding and hence an important sanction that can be used as a means of control in the public sector is not available in relation to companies. In addition, it is conceivable that corporations that are controlled by strict regulations in their own countries will attempt to conduct their research in jurisdictions that do not require compliance with guidelines. That possibility is of particular concern in the sense that escaped pathogens do not respect national borders. Accordingly, the discussion in this book will extend from national to international mechanisms for legal control.

Despite the fact that attempts to control science through the use of law strike at the cherished concept of freedom of scientific inquiry, legislation for the purpose of controlling genetic engineering may be required. It could provide the answer to the complex problems that are inherent in controlling a technology that involves substantial benefits and substantial risks. A discussion of liability, compensation, and legal control with reference to analogous statutes in force in New Zealand and overseas will reveal that those difficulties are not insurmountable. The measures that will be considered include the licensing of scientists and laboratories and the establishment of a special committee that would evaluate proposed uses and experiments. Freedom of information, patentability, the special problems of corporate control, and the potential role of health departments and courts in enforcing such legislation will also be examined, as will important legal issues that arise under existing statutes and at common law.

---

*As defined in Chapter 1 of this book.

NOTES

1. It has also been suggested that a recent outbreak of smallpox, which resulted in the death of Janet Parker, a Birmingham University photographer, was attributable to recombinant DNA experimentation at the laboratory from which the virus escaped. The research at the laboratory was centered on the genetic analysis of human and animal pox viruses and the comparison of biochemical changes in infected cells. Studies on hybrid smallpox-cowpox strains were also in progress. Although DNA mapping tests have not yet been completed, these factors lend support to the view that Parker's death was caused by a recombinant virus. See "Was the Birmingham Outbreak Really Smallpox?" *New Scientist,* October 19, 1978, p. 155.

2. "Recommendations of the MRC Advisory Committee on Genetic Manipulation" (as amended by the DSIR), 1977 (unpublished).

3. For example, N.Z. Patent Application No. 163521, "Improvements In Or Relating To Micro-organisms," filed by Ranks, Hovis, and McDougall Ltd. on April 30, 1971, and N.Z. Patent Application No. 187300, "Recombinant DNA Transfer Vector and Micro-organism Containing a Gene From a Higher Organism," filed by the Regents of the University of California on May 17, 1978. In the United Kingdom a patent has been granted to the General Electric Company in relation to a genetically engineered organism and a genetic engineering technique, see Patent Specification 1,436,573 (U.K.). See also "How Genentech Made Human Insulin," *New Scientist,* September 28, 1978, p. 926.

CONTROLLING TECHNOLOGY

1

THE TECHNOLOGY OF
GENETIC ENGINEERING

THE FOUNDATIONS

Since Van Leeuwenhoek perfected the simple light microscope in the seventeenth century, knowledge of intracellular components has increased with mitotic impetus. Yet in a different era scientists worked without the compelling visual evidence of structures like the dividing chromosome. In the fourth and fifth centuries B.C., Hippocrates and Aristotle postulated theories of reproduction and inheritance that had already been referred to, in theological rather than scientific terms, in the Old Testament and in the writings of Homer in the eighth century B.C. In the *Iliad* considerable attention was devoted to the lineage of the different families that were involved and to the hero's inherited characteristics. Similarly, in the *Odyssey* some emphasis was placed on the characteristics that might flow from a "noble birth."[1] By the seventh century B.C., descriptions of practical eugenic applications of these theories had been recorded by another Greek writer, Hesiod. This was paralleled in the *Republic*, where Plato discussed eugenic laws that would help to ensure the maintenance and preservation of the ideal state.

A scientific age that manifested itself in the empirical observations of Hippocrates and Aristotle had also marked the foundation of the atomist school by Leucippus and Democritus in the fifth century B.C. They hypothesized that all organisms consisted of an infinite number of minute

1

particles of matter known as atoms. The proponents of this theory were of the view that generative material was of similar composition and that it did not come from any particular site in the body. These were important developments in the sense that they marked the acceptance of a material rather than a spiritual basis for the transmission of genetic factors. Such theories and their speculative off-shoots persisted. They became the subject of careful consideration by nineteenth-century geneticists like Charles Darwin nearly 2,300 years later. In fact Darwin explicitly compared Hippocrates's concepts of heredity with his own discredited "provisional hypothesis of pangenesis" and the Hippocratic doctrine is now referred to as pangenesis.[2]

The contributions of the early Greek scientists were not matched by their Roman counterparts. Similarly, the more distinguished naturalists of the medieval period such as Albertus Magnus, Thomas Aquinas, and Roger Bacon were content to accept Hippocratic or Aristotelian tenets, subject to minor disagreements over peripheral points.

Scientific knowledge in this area had advanced little further by the end of the sixteenth century. Even Leonardo da Vinci, who described the Aristotelian view, did not criticize it. In fact it was not until the seventeenth century, and the development of an efficient microscope, that there was a significant increase in the understanding of reproductive processes.

A period of intense interest in hybridization experiments followed in the eighteenth and early nineteenth centuries. The most notable of these studies were conducted by Kölreuter and Gärtner, whose work was continued by Mendel in the latter half of the nineteenth century. Mendel was arguably the first person to explain the scientific basis of selective agricultural techniques that had existed since approximately 5000 B.C. The factors that distinguished Mendel's work from that of his predecessors can best be described in his own words. In 1865 he wrote:

> Those who survey the work done in this field will conclude that of all the numerous experiments made, not one has been undertaken to such an extent and in such a way for it to be possible to determine the number of different forms in which the offspring of hybrids appear, or to classify these forms with certainty according to their separate generations, or to ascertain their statistical relations. Courage is indeed required to undertake such extensive labours. But they would seem to constitute the only right way of ultimately achieving the solution to a question which is of inestimable importance in connection with the evolutionary history of organic forms.[3]

The scientific community of the time failed to understand the significance of Mendel's work and he was largely ignored. Ironically, however, the "evolutionary history" of which he spoke had already been considered and had captured the attention of the more influential thinkers of the day. Charles Darwin, writing in the wake of Lamarck, published *Origin of Species by Means of Natural Selection* in 1859. Further publications followed but none of the references to the extensive hybridization literature includes mention of Mendel. In fact there is some doubt as to whether Darwin would have appreciated the importance of Mendel's studies even if he had known of them.[4]

Mendel had read Darwin's *Origin of Species*, although he did not refer to it in any of his papers. He may have wished to avoid the controversy surrounding Darwin's theory of evolution, although involvement in that debate might have promoted an increased understanding of his own work. Clearly there would have been points of disagreement between the two in terms of hereditary variability, but comparison could conceivably have benefitted both.

Galton was not as reticent as Mendel. He criticized much of what Darwin had written. However, his fame was not merely as a critic. His "Law of Ancestral Heredity" described certain resemblances between parents and offspring but he too failed to put forward an underlying Mendelian type of principle that would have explained his observations. He was probably the first person to systematically study human genetics and was commonly regarded as the founder of the science of eugenics. His idea on eugenics had crystallized after reading *Origin of Species* — in his view human evolution could be guided by substituting social control for natural selection. He also acknowledged that such considerations imposed new moral duties on those who had the power to alter the shape of society.

The work of some of Darwin's most important critics, including Weismann, Bateson, and de Vries, was principally responsible for the belated recognition of Mendel's contribution, which came in 1900. It was the signal for the dawning of the new age of genetics. At the end of the nineteenth century the study of heredity and variation was an acknowledged but not an established science. Bateson helped to remedy this by introducing the term "genetics" in 1906 and founding the Cambridge School of Genetics in 1908. At the third International Conference on Genetics in 1906, Bateson presented a paper entitled "The Progress of Genetic Research." He wrote:

Like other new crafts, we have been compelled to adopt a terminology which, if somewhat deterrent to the novice, is so necessary a tool to the craftsman that it must be endured. But though these attributes of scientific activity are in evidence, the science itself is still nameless, and we can only describe our pursuit by cumbrous and often misleading periphrasis. To meet this difficulty I suggest, for the consideration of this congress the term *Genetics*, which sufficiently indicates that our labours are devoted to the elucidation of the phenomena of heredity and variation: in other words, to the physiology of descent, with implied bearing on the theoretical problems of the evolutionist and the systematist, the application to the practical

problems of breeders, whether of animals or plants. After more or less un-
directed wandering we thus have a definite aim in view.[5]

The word "genetics" was accepted by the scientific community as were
many other related terms that were proposed by Bateson. Other researchers
began to coin words and the "gene" was first referred to in 1908 by Johann-
sen. Chromosomes had already been named by cell biologists in the late
nineteenth century and it now remained for eminent cytologists like W. S.
Sutton and E. B. Wilson to formulate a chromosome theory of heredity.
Their work was continued by T. H. Morgan whose *Drosophila* studies
finally established genetics as a fundamentally important life science.[6]

In the USSR the years that followed Morgan's discoveries bore witness
to a disturbing blot on the influence of Marxian and Leninist thought on the
development of science. "Morganism," as it was subsequently labelled by
the state, came to be viewed as the nonproductive embodiment of a reac-
tionary and idealistic approach to science. Müller, one of Morgan's more
renowned students, represented the Morganist school. His colleague,
Lysenko, had propounded the so-called Michurin approach, which was
equivalent to the discredited Lamarckian stance. He denied the clear scientific evidence of the Mendelian, Morganist school, and his argument, which
was initially couched in the language of genetics, degenerated into a
political struggle that culminated in Müller's departure from the USSR. On
July 31, 1948, Lysenko delivered an address to the Session of the Leni.
Academy of Agricultural Sciences of the USSR. The members of the
academy immediately adopted a resolution on "the situation in biological
science." They wrote:

> Scientific research in a number of biological institutes, and the teaching of
> genetics, plant breeding, seed cultivation, general biology and Darwinism
> in Universities and Colleges, is based on syllabuses and plans that are
> permeated with the ideas of Mendelism-Morganism, which is gravely pre-
> judicial to the ideological training of our cadres. In view of this, this general
> meeting is of the opinion that scientific research in the field of biology must
> be radically reorganised and that the biological sections of the syllabuses of
> educational institutions must be revised.[7]

Soviet biology has not yet recovered from the harm that resulted from
this "reorganization." Predictably enough, Lysenko and his school were
eventually discredited, but the developments that were already beginning to
take place in England and the United States would not reach the USSR for
some time to come.

A new era of molecular biology commenced in the West with attempts
to understand the nature and function of the gene. In 1944 Avery,
MacLeod, and McCarty demonstrated that deoxyribonucleic acid (DNA)

was the carrier of a virulence trait in the bacterium with which they were working. By the early 1950s Hershey and Chase had confirmed the tentative Avery hypotheses and expanded upon them. It was now clear that DNA was capable of transmitting inherited characteristics. The way in which DNA functioned was not yet known and would depend on an understanding of the structure of the DNA molecule.

The first insights were provided when Wilkins and Franklin began to apply X-ray diffraction techniques to the study of DNA structure.[8] Pauling, who pioneered the use of X-ray diffraction data to test his models of molecular structure, promised to produce a model of the DNA molecule in the near future. These developments no doubt added impetus to the work of Watson and Crick, who published their results in April 1953. They constructed a model of the DNA molecule and explained the process by which DNA was replicated within the cell. The gene could now be viewed as a site on a large number of base points in a chain of nucleotides that linked the strands of the double-helix.

Throughout the 1950s and 1960s genetic science advanced at an unprecedented rate. For example, in 1961 Nirenberg matched certain components of the DNA molecule with gene function and the knowledge that was gained about the mechanisms of heredity led the way for the refinement of experimental techniques in somatic cell genetics. In the 1970s advances in the fields of virology and enzymology, which were largely pioneered by Berg, Boyer, Cohen, Jackson, Hogness, and Nathans resulted in the birth of recombinant DNA research.

Once born the science quickly gained respectability — a respectability that has already been recognized in the form of the 1978 Nobel Prize for medicine, which was awarded to Nathans, Smith, and Arber for their work with restriction enzymes that cleave DNA at specific sites.[9]

THE NEW TECHNOLOGY

Because the terminology of the "new craft" to which Bateson referred has been partly superseded by the language of the new technology of genetic engineering, a brief description of some of the relevant terms and techniques is necessary. The scope of this book has been largely predetermined by the Report of the New Zealand Working Party on Novel Genetic Techniques.[10] Recombinant DNA research, protoplast fusion, and somatic cell genetics were reviewed in the report. Those techniques form the focus of this book. They share many features including the capacity to alter the genetic composition of plant and animal cells. The possibility that they may cause unique physical damage to the environment is another common factor.* The

---

*Cloning and gene synthesis techniques raise ethical issues but, particularly in the former case, they do not, in themselves, give rise to serious physical dangers.

working party also referred to the wider category of "microbiological hazard" that encompasses the techniques described above and necessitates consideration of the risks inherent, for example, in virology research.

In its broadest sense, recombinant DNA research "involves combining molecules of different biological origin by any methods that overcome natural barriers in mating and recombination, to yield molecules that can be propagated in some host cell, and the subsequent study of such molecules."[11] However, a narrower and more useful definition in terms of *in vitro* recombinant DNA research, which appears to have gained widespread acceptance, is as follows: "*In vitro* recombinant DNA research is the use of restriction enzymes or other biochemical methods to prepare fragments of DNA that are subsequently combined to yield molecules that can be propagated in a host cell."[12]

In vitro research is conducted outside the organism and is the opposite of *in vivo* research, which is considered to be less dangerous, principally because of *in vivo* limitations relating to the natural resistance of the host organism to the survival, replication, or expression of novel genetic material.

As the last quotation indicates, the recombinant DNA technique relies on the precise action of restriction enzymes. These enzymes are commonly used to cleave DNA at particular sites. Each enzyme is extremely specific and operates only on certain of the base sequences of the DNA structure. After cleavage a segment of DNA from a different organism is inserted and a ligase, which is the counterpart of the restriction endonuclease, is used to anneal (or rejoin) the bonds that have been broken.

After the DNA has been recombined the next step in the procedure is to induce it to replicate within a host cell. Plasmids have proved to be extremely effective vectors or carriers of recombinant DNA. They consist of genetic material that leads a stable existence in a bacterial cell independent from the preexisting chromosome.[13] Hence they are almost ideal vehicles for the transmission of recombinant DNA molecules into host bacterial cells where it is hoped that they will replicate.

Viruses (or phages) that infect bacteria (bacteriophages) have also assumed increasing significance as vectors. Because viruses are essentially units of DNA or RNA (ribonucleic acid) surrounded by protein they provide a very efficient means of transferring recombinant DNA into host organisms. They also multiply within the bacterial host, thus producing clones of the recombinant DNA.

The host organism is generally the *Eschericia coli* bacterium. This microorganism exists, in its most common form, in the human intestine. The K-12 strain, which is the laboratory variant of this bacterium, is not normally found in the gut, although it could survive there. It is primarily in this context that the "shotgun" experiment has given rise to a considerable amount of concern. Such an experiment could involve the introduction of numerous unselected fragments of DNA into an *E. coli* bacterium. Interesting or potentially useful DNA sequences would then be isolated from the resultant mixture.

Protoplast fusion experiments also result in the formation of new genetic material. In fact a new cell with altered cytoplasmic, or extranuclear fluid, is produced. The technique that is used differs from the one employed in classic recombinant DNA studies. Entire cells are induced to merge by the use of procedures that by-pass normal sexual processes. For example, carrot, mouse, or bacterial cells can be fused with human cancer cells.

Both in this context and in relation to recombinant DNA research a distinction between prokaryotic and eukaryotic cells must be recognized. Prokaryotic cells are "primitive" cells that have neither an organized nucleus nor a nuclear membrane that surrounds the chromosomes—bacterial cells fall within this category. On the other hand eukaryotic cells are advanced cells with a highly organized nucleus and a nuclear membrane. The cells of all advanced organisms are eukaryotic. When recombinant DNA or protoplast fusion research is conducted the novel genetic material that is produced can be composed of both eukaryotic and prokaryotic DNA material. Thus the techniques can be used to circumvent the natural evolutionary barriers that exist in terms of constraints on interspecies sexual reproduction.

In a broad sense the third area of interest, somatic cell genetic research, relates to genetic phenomena that occur in somatic cells as opposed to reproductive or germ cells—the somatic cell contains the organism's full complement of paired chromosomes, whereas the chromosomes in the reproductive cells are unpaired. Hence somatic cell genetic studies can involve protoplast fusion to the extent that somatic cells are merged. Conversely, this type of experiment need not involve two cells but can, for example, be limited to the alteration of the genetic composition of a single cell by mutagenesis or induced mutation.

In terms of control the distinction between physical and biological containment is of particular importance. Physical containment procedures involve attempts to prevent experimental organisms from escaping from the laboratory. On the other hand, biological containment is dependent on the production of organisms that cannot survive or reproduce outside the laboratory. This is usually achieved by inducing successive mutations that disable the organism so that it can grow and reproduce only in an unnatural environment.

NOTES

1. Hans Stubbe, *History of Genetics from Prehistoric Times to the Rediscovery of Mendel's Laws* (Cambridge, Mass.: MIT Press, 1972), p. 14.

2. Charles Darwin, *The Variation of Animals and Plants under Domestication* (London: J. Murray, 1868).

3. Gregor Mendel, *Experiments in Plant Hybridization* (Cambridge, Mass.: Harvard University Press, 1938), p. 3.

4. Cyril D. Darlington, *Darwin's Place in History* (Oxford: Blackwell, 1959), pp. 51-56.

5. Cited in Stubbe, op. cit., p. 272.

6. Thomas H. Morgan, *The Theory of the Gene* (New Haven, Conn.: Yale University Press, 1926).

7. *The Situation in Biological Science.* Proceedings of the Lenin Academy of Agricultural Sciences of the USSR, July 31-August 7, 1948 (Moscow: Foreign Languages Publishing House, 1949), p. 71.

8. These techniques are described in Garland E. Allen, *Life Science in the Twentieth Century* (New York: John Wiley, 1975), p. 219.

9. See "Gene Manipulators' Tool Wins the Nobel Prize," *New Scientist,* October 1978, p. 156.

10. "Report of the Working Party on Novel Genetic Techniques," April 1978, G.21A. The working party was established in July 1977 by the minister of Science and Technology.

11. See "Report and Recommendations of the European Molecular Biology Organisation Standing Advisory Committee on Recombinant DNA," Second Meeting, 1976, p. 3.

12. Ibid. This definition was cited with approval in the Medical Research Council Guidelines. See "Recommendations of the MRC Advisory Committee on Genetic Manipulation," 1977 (unpublished), p. 2.

13. Peter L. Bergquist, "The Uses and Abuses of Recombinant DNA Molecules, Aspects of their Molecular Genetics and Molecular Biology," 1976 (unpublished).

2

LAW AND SCIENCE: SOME PERSPECTIVES ON CONTROL

SHOULD SCIENCE BE CONTROLLED?

Genetic engineering research can be viewed as part of a wider quest for knowledge that pervades all scientific endeavor. To control science is arguably to deprive the community of what has come to be viewed as its right to truth and knowledge. On the other hand it can be suggested that there is a distinction between controlling science, or a science, and controlling technology. Hence freedom of inquiry might be justified by reference to the value of knowledge for its own sake or to the value of knowledge for the sake of its applications. In pre-Baconian times such a distinction was more meaningful than it is today and although the dichotomy is valid it tends to produce an easy but overly simplistic solution to the important problems of control and progress. Namely, that it is technology and not science that should be controlled.

But what is progress and why should it not be impeded? Teilhard de Chardin has argued that progress is the goal of the universe.[1] Karl Popper adopted a similar stance when he stated that knowledge is ever-growing.[2] Perhaps control is a concept that is inimical to the essence of science. If science is "ever-growing" how can it be controlled? Clearly control is not necessarily synonymous with a complete ban nor does it necessarily imply control imposed by "outsiders" on science and scientists. For instance, it has been argued that the methodology of science constitutes its own inbuilt control. In the course of considering Bronowski's views on these issues Sutton states:

> Bronowski discussed why it is that science has had such a profound moral and social influence. Obviously it is not because individual scientists are especially virtuous, or any less moved by emotions and opinions than other men and women. Still less is it because of some code of conduct imposed on scientists by society at large. Rather it is because certain values are imposed *by the scientific method itself* for unless scientists are open to new ideas and interpretations, and are sceptical yet willing to accept the integrity of others, then the scientific method can make no progress.[3]

Bronowski goes further than Sutton suggests. He says that the independence of the scientist must be protected. He emphasizes a scientific society "where men are committed to explore the truth."[4] In fact Bronowski has proposed the "disestablishment" of science.[5] This would ensure the fullest possible separation between science and government in order to free science from political control. It would remove the choice of priorities in research from the hands of governments, but before the ties between science and government could be severed it would, as Bronowski points out, be necessary to minimize the pressure that governments can exert through the manipulation of research grants and contracts. This, he argues, could be achieved by providing for a single, overall grant for research that would be divided by all the scientists in a particular country.[6] In the long term he envisaged an international fund.

Bronowski's ideas appear to have been based on a very firm belief in the integrity and judgment of scientists, and it is interesting to compare his attitude with that of C. P. Snow. He wished to limit the power of individual scientists who were involved in governmental decision-making processes. Snow stated that "the nearer he [the scientist] is to the physical presence of his own gadget, the worse his judgment is going to be."[7] He recognized that technical expertise is indispensable in an advisory context and was against what he described as the secret choice, but he was inclined to value the foresight rather than the integrity of the scientist in the decision-making arena.

Society might be best served by a compromise between participation by the scientific community in government and watchful governmental control of science. In particular, attempts to sever the ties between science and government may be misguided. While opinions on the role of scientists as controllers of technology will differ, it is clear that the scientist must share the responsibility for the way in which his discoveries are used. Perhaps he is the new Prometheus with a twofold burden as scientist and citizen. The scientist, as scientist, is a professional who, in his search for knowledge, is motivated not only by the public welfare but by self-interest as represented by career advancement and public acclaim. The latter may, in itself, operate as a control mechanism, although the record of abandonment of projects because of individual scientists' moral reservations about their consequences has hardly been encouraging.[8] Wider public disapproval appears to be necessary. Thus if the consequences of a discovery are likely to be viewed as harmful and undesirable, the scientist might be reluctant to proceed. However, such reluctance will depend on the state of knowledge of both the scientist and the public regarding the effects of the research. In this context Marcuse has referred to a tendency for science and technology to become the instrumentalities of those who simply want to perfect techniques without addressing themselves to wider questions of progress in terms of the consequences of their research and the question of who it is designed to benefit.[9]

In examining these issues a risk/benefit analysis is in theory, if not in practice, of fundamental importance. The analysis helps to provide an indication of whether the technology should be prohibited or controlled until it is proved to be safe. It involves the weighing of the harm that may flow from a technology against the benefits that are inherent in it. On the other hand, that phraseology introduces a certain bias that can arguably be justified on the basis that although the risk often appears to be distant, speculative, and imprecisely defined, the benefits are often clear and present. In addition, a discovery may fulfill a pressing need, as in the case of the production of nuclear fuel. Even when a harmful effect is known to result from the application of a particular technique or product, the benefit may be seen to outweigh harm that has assumed the proportions of a certainty rather than a risk. For example, cytotoxic cancer drugs are in common use even though they are known not to be perfectly specific.

The interrelationship between science, government, the scientist, and the citizen is complex and dynamic. Numerous variables can influence or upset the delicate network of checks and balances that operate between the citizen, the politician, the technician, and the technique. The series of events that led to the creation and deployment of the atomic bomb serves to illustrate some of these interactions.[10] In that context, as in many others, the important variable was lack of information. That was a significant factor not only in terms of the scientific community's lack of understanding of the physical consequences of atomic technology but also because scientists in Britain and the United States were under a misapprehension as to the state of knowledge of their German counterparts. Although a certain amount of wartime secrecy was necessary, the public and the vast majority of those employed on the project were informed neither of the details of what was under construction nor of its implications insofar as they were known at the time. The public was not given a chance to debate the issue and the responsibility for controlling the technology did not rest with the scientific community. Perhaps the scientists fulfilled their duty when they informed the government of the possible implications of their research. If the scientist is seen to have a wider duty in terms of voluntary self-censorship then this important role is not acknowledged, in any practical sense, by the scientists themselves.[11]

The scientist has no special expertise in matters of ethics or morality. In that sense he is in the same position as the vast majority of citizens who would be affected by his decision. In balancing risks against benefits his expertise is in the purely technical field in terms of physical cause and effect. In New Zealand virtually no emphasis is placed on conducting ethics-related courses for scientists. Despite discussions by the New Zealand Association of Scientists, funds are not set aside for this purpose.[12] In other countries the situation is little better.[13]

Scientists are often criticized as decision makers on the basis that they are ill-equipped to make decisions that involve ethical factors; but are nonscientists equipped to make decisions that involve scientific factors? Some would suggest that if science is to be controlled it must be by a technocracy. Such an argument proceeds on the basis that only those who fully understand scientific techniques are in a position to make decisions concerning their control. This is patently untrue. Less-than-expert knowledge may be sufficient to warn of unacceptable risks and it is the very question of acceptability that cannot be left to the scientists. But just as scientists should be required to consider ethical matters, so also must nonscientists be educated in the scientific facts. Presumably this will help to ensure that an informed decision, based on the widest societal considerations, is taken.[14]

The scientist may be a philosopher and a philanthropist but he is not an elected representative of the people. His search for knowledge, even if unadulterated by less worthy motives, cannot be allowed to assume a position of inviolable primacy. Like every other citizen the scientist must bow to the wishes of the majority. However, that analysis presupposes that there is a free flow of information from the scientist to the citizen either directly or via government. The wider community must have information if it is to entrust the decision-making responsibility in particular cases to its elected representatives and reserve to itself the right to influence decisions either by public debate or by the sanction of the ballot box. The development of startling new technologies, whether atomic or biological, has highlighted these issues. Perhaps the lessons of the past will help to ensure that in the future the control of new technologies is placed firmly within the grasp of those who will be most affected by them.

SHOULD GENETIC ENGINEERING BE CONTROLLED?

At a very simplistic level it can be argued that the new technology of genetic engineering should be controlled in New Zealand because it has

already been controlled overseas.[15] This argument is based on several as-
sumptions. It presupposes that the existence of regulations in other countries
will encourage overseas interests to conduct their research in New Zealand*
and that that would be undesirable largely because the technology is
perceived to involve unacceptable dangers. The accuracy of these proposi-
tions can best be assessed by means of a risk/benefit analysis. However,
even that type of inquiry is of limited assistance in this context. Like the
nuclear physicists of the 1930s and 1940s the genetic engineers of the 1970s
do not know precisely what is involved in their research.

The benefits that are promised by genetic engineering techniques are
clear and spectacular. They include cancer cures,[16] increased understanding
of gene function and the amelioration of genetic defect, the development of
improved antipollution and agricultural techniques, the bulk production of
scarce antibodies and hormones,[17] and the manufacture of new industrial
chemicals. Some of these breakthroughs have already been achieved. For example, scientists have successfully produced somatostatin by the use of
genetic engineering techniques. At present this hormone costs $50,000 a gram
in the United States. Genentech, a company based in California, will market
the hormone at a substantially reduced cost.[18] Its researchers have also used
engineered genes to convert bacteria into insulin-producing units.[19]

Dangers may also be inherent in the new developments. For instance,
Wald states that "if right now I had to weigh the probabilities of either event I
would guess that recombinant DNA research carries more and earlier risks of
causing cancers than hope of curing them."[20] Chakrabarty, a microbiologist
who is employed by the General Electric Company, has developed a strain of
pseudomonas bacteria that can digest crude oil. He acknowledges the dangers
that are associated with his invention and admits that it may not be possible
to control the organism in such a way as to limit its consumptive capacity to
oil spills. It could conceivably attack oil in the ground or in the refinery.
Chakrabarty also points out that while it is now possible to manipulate *E. coli*
bacteria and to produce a strain that converts cellulose into assimilable sugars
and fatty acids, this seemingly desirable capability has resulted in increased
production of fatty acids and sugars at a rate that exceeds the absorptive
capacity of the human intestine. This has led to dietary problems and the formation of harmful toxins.[21]

Stanley Cohen has expressed the view that the risks that are inherent in
the research are potential while the benefits are actual. He states:

> For all but a very few experiments, the risks of recombinant DNA research
> are speculative. Are the benefits equally speculative or is there some factual

---

*It is interesting to note that U.S. corporations have already attempted to avoid stringent
interpretations of patent laws in their own country by applying for patents in New Zealand.

> basis for expecting that benefits will occur from this technique? I believe that the anticipation of benefits has a substantial basis in fact, and that the benefits fall into two principal categories: (i) advancement of fundamental scientific and medical knowledge, and (ii) possible practical applications.[22]

Cohen succeeds in drawing the distinction between science and technology but has little concrete to offer in terms of an assessment of the magnitude of relevant risks and benefits. He is obviously of the opinion that the scientific community must, if possible, advance statements of fact to support their conclusions on the control of the research and it is clear that if scientists cannot discuss "the facts," their opinions may have to give way to informed community views. For example, in March 1978 researchers were warned by a member of the U.S. House of Representatives that "they can no longer presume to tell the rest of the people that they are ignoramuses who lack the ability to make intelligent judgments about things you think we know nothing of." He continued:

> I would urge that the scientific community as a whole begin to think more seriously of itself as only a group of citizens in our community of citizens. The debate over recombinant DNA research has given me a fascinating insight into the arrogance and anti-democratic spirit of which the "established" scientific community has shown itself capable. Those who have lobbied so hard to prevent the Congress from taking steps many of us are convinced need to be taken to protect the public health and safety have shown that they apparently think themselves omniscient and infallible.[23]

This concern has been echoed by many scientists. In fact it has been claimed that the proponents of genetic engineering research are attempting to suppress dissent and intimidate their opponents.[24] The jobs, grants, theses, and employment prospects of many of the critics have apparently been threatened.

However, even if members of the community are presented with all the available scientific evidence, it is likely that public opinion will be just as divided as scientific opinion. If Nobel prize winners like Lederberg[25] and Wald cannot agree, there is no reason to suspect that an obvious community consensus will be discernable. Those who oppose genetic engineering argue that techniques that involve the creation and use of new forms of life ignore the natural evolutionary process that has a 3-billion-year history. Wald points out that it has taken from 4 to 20 million years for a single mutation in one amino acid in the haemoglobin sequence to establish itself as a species norm. That can be compared with the laboratory transposition of whole proteins into new associations and with interspecies transfers of genetic material between prokaryotic (primitive) and eukaryotic (advanced) organisms.[26] For example, carrot and tobacco cells have been fused with human cells and grown in culture.[27]

On the other hand it is important to note that DNA recombinations can take place naturally. For instance, it has been demonstrated that the *E. coli* bacterium is exposed to free human DNA that is released from broken cells in the gut. Under those conditions it is conceivable that random recombination may have been occurring continuously.[28] In addition, such combinations are no doubt tested by natural selection, which may ensure that unusual associations will neither be expressed nor replicate in the host organism.[29]

These factors are by no means certain and in the experimental context it is quite feasible that trials involving billions of microorganisms could produce a novel organism that is capable of surviving and reproducing outside the laboratory. The probability of damage is increased by the nature of the experimental process. For instance, so little is known about gene function that a "shotgun" technique is commonly used. It involves the random insertion of different types of foreign DNA into a host organism. After the recombination the researcher examines the results and selects out interesting organisms for study. This type of random experiment involves a considerable magnification of the risks that are associated with recombinant DNA research.* Another worrying factor that should not be overlooked in this context is that many of these experiments are carried out on the *E. coli* bacterium. In its most common form, this bacterium inhabits the human gut. A weakened K-12 strain is generally used as the host but the possibility of the survival of that strain in humans is not denied.[30]

A more subtle argument that runs parallel to the evolutionary one is that this type of research prima facie alters the natural balance of the host organism and its environment in unpredictable ways. Even if seemingly desirable new characteristics are expressed in a host organism, unseen implications may be present. For example, the hormonal balance of the organism could be disturbed and its wider function in the biological community could be impaired.

There is also an important distinction between *in vivo* and *in vitro* experiments. Many of those who acknowledge that genetic engineering should be controlled would confine regulation to *in vitro* experiments on the basis that in the case of *in vivo* research the limitations that are imposed by the

---

*For example, if certain foreign genes are expressed in the host cell this would result in the formation of a protein foreign to the infected cell or the uncontrolled synthesis of a normal protein. The likelihood of hazard would be significantly reduced if purified DNA that has been characterized as nonhazardous were used instead of random fragments representing a wide range of genetic information.

organism itself will be effective in preventing the expression of harmful characteristics. However, it may be unwise to discard the risks involved in *in vivo* research and it is at least arguable that the likelihood of the expression and replication of undesirable combinations is sufficient to justify regulation of the *in vivo* technique.

Control can take the form of a complete or temporary ban on some or all types of experiments or the specification of regulations for permissible experiments. Those who advocate regulation freely admit that there are risks inherent in the research but argue that these risks can be reduced to an acceptable level. The Royal Commission on nuclear power generation in New Zealand has considered this argument in a different context. The commission noted that

> no technology (including any kind of electric power generation) is absolutely safe. Risk of death or injury is a price of existence. Modern technological society tries to reduce the risk to what it considers to be acceptable. At this level the risks are assumed to be less than the advantages, which implies a subjective evaluation of what is an acceptable risk.[31]

The risks of genetic engineering can be reduced by means of biological or physical containment or both but how reliable are these containment measures? Available accident statistics are less than impressive. At Fort Detrick, the highest level of containment laboratory in the United States, there have been 423 accidental infections and three deaths over a 25-year period—on average one accident every three to four weeks. Most of these accidents were apparently due to noncompliance with safety regulations.[32] At Porton Down in England, which is reputed to be the highest level containment facility in Western Europe, there are also recorded instances of infection. For example, one research worker was infected with viral haemorrhagic fever as a result of the accidental penetration of protective gloves. There is no known cure for this disease. None of the infections at Porton Down or Fort Detrick appears to have resulted in epidemics in the surrounding communities,[33] and to that extent they are similar to the accident that caused the death of the *pinus radiata* seedlings at the DSIR Plant Physiology Laboratory at Palmerston North.

On the other hand recent viral escapes reinforce the view that the risk of epidemic is not small enough to be discounted. A smallpox outbreak that spread from a London laboratory in 1973 resulted in the death of the carrier and two people who had never entered the laboratory.* In late August 1978 an even more disturbing incident occurred in Birmingham in spite of the fact

---

*For a full description, see Chapter 3, section on "the United Kingdom."

that a Committee of Inquiry had produced an extensive report on the events in London and had formulated stringent safety precautions that were voluntarily adopted by most viral laboratories in the United Kingdom.[34]

Clearly then, it would seem to be almost impossible to completely prevent accidents. In terms of physical containment there are problems involved in keeping engineered organisms within the confines of the institution in which they are used and keeping normal organisms out. For example, if bacteria entered a laboratory, disabled organisms could transfer their recombinant DNA to the normal bacteria, which might then escape and replicate in the outside environment. Hence the biological containment back-up is at least partly dependent on the success of physical containment measures.

Awareness of these risks has resulted in the development of experiments designed to determine the probabilities and processes of accident, infection, expression, and spread. These so-called scare scenarios are more important than their label would suggest. They are currently being conducted at Porton Down and Fort Detrick and it is hoped that they will provide information about the risks that are attached to the different categories of experiment.[35] The insights gained from this work could conceivably form the basis of more efficient methods of control.

Genetic engineering techniques may also have important implications for society at quite a different level. For example, agricultural benefits that could manifest themselves in the form of increased productivity are closely linked with the country's economy. As the techniques continue to develop, scientists are prepared to acknowledge these implications. In August 1977 Sutton commented that genetic engineering techniques had not "demonstrated probability of any socioeconomic import at all."[36] By March 1978 he had come to the conclusion that within a few months his research might lead to discoveries that could save New Zealand farmers millions of dollars in fertilizers.[37]

Economic implications arouse political interest and may lead to political interference. The effects of governmental involvement in crop genetic research in the USSR in the 1940s* still remain today as a warning to those who would ignore inconvenient scientific evidence in order to further economic goals.[38] It is also conceivable that governments will develop an interest in using genetic engineering as a means of ordering society,[39] and in this context the brave new world is distant but daunting. If Bronowski's faith in the integrity of science and scientists proves to have been misplaced, the wider community can still rely on attempts to control the technologists. However, it is somewhat more difficult to control governments. Full

---

*See Chapter 1.

information is needed for this important purpose because, as C. S. Lewis points out, "if any one age really attains by eugenics and scientific manipulation, the power to make the descendants what it pleases, all men who live after it are patients of that power. They are weaker not stronger."[40]

In conclusion, it can be stated that the technology of genetic engineering should be controlled. A complete ban on genetic engineering activities could be regarded as an unnecessarily insular restraint on scientific inquiry and the benefits that it offers.[41] On the other hand genetic engineering techniques involve obvious risks. In order to maximize benefit and minimize risk, thoughtful regulation is imperative. Choices must be made and risks evaluated in terms of the source of the genetic material, the nature of the vectors and host organisms, and the procedures and quantities involved. The community should not delay action until after there has been a dramatic physical manifestation of the dangers that are inherent in the technology.

METHODS OF CONTROL

Informal and formal means of controlling genetic engineering can be envisaged. Informal controls include peer group pressure and the withdrawal of funds by sponsoring agencies. In terms of a more formal method of control, it is possible to draft nonmandatory guidelines that relate to the conduct of the research. Such guidelines would be backed only by funding constraints and peer pressure.[42] The final possibility is legal control. It could take the form of a new statute relating solely to the technology of genetic engineering or the introduction of a statute concerned with the control of wider categories of microbiological hazard. The utilization of existing common law and statutory remedies must also be considered,* in addition to the promulgation of a set of technical regulations under a new or an existing statute.[43]

As far as legal control is concerned, the options are relatively clear. Unfortunately, the same cannot be said of the solution to problems concerning the best possible means of legal control and to the prior question of whether legal intervention is preferable to informal and formal nonlegal controls. Those issues will be considered on the basis of an evaluation of existing common law and statutory remedies† and after an examination of attempts to control genetic engineering.‡ However, at this point an outline of the

---

*See Chapter 4, section on "Existing Law."
†*See Chapter 4, first section, for a discussion of the need for legislative action.
‡See Chapter 3.

various sources of the legal sanction is insufficient. It is necessary to refer briefly to the possible content of laws that could be used to control the new technology. The alternatives include the introduction of a complete or partial ban on genetic engineering, the implementation of controls on the manner in which experiments are conducted (plus a total absence of prohibitions on particular categories), or the adoption of a combined approach involving the prohibition of certain specified experiments and the control of others.

Effective enforcement of each of these regimes would necessarily involve an inspectorate. The second and third alternatives would require compliance with prescribed physical or biological containment procedures (or both) and could conceivably result in the establishment or utilization of statutory committees. These committees might be responsible for issuing genetic engineering licences in relation to projects, institutes, or individual project supervisors.*

Provision could also be made for injunctive remedies and emergency procedures. In addition, an efficient system of control would ensure that preventive measures were backed with compensatory remedies. Thus the recovery of losses caused by the technology would be facilitated. However, the ultimate legal sanction is provided by the criminal law and for that reason any attempt to control genetic engineering by legislation is likely to involve offense and penalty provisions.

In the following chapter a review of the events that have taken place in countries that are moving toward the implementation of controls on genetic engineering will serve to illustrate the nature of the choices that are available and the latent difficulties that may be involved in the various methods of controlling this technology.

NOTES

1. Pierre Teilhard de Chardin, *The Phenomenon of Man* (New York: Harper, 1959).

2. Karl Popper, *The Logic of Scientific Discovery* (London: Hutchinson, 1968).

3. Bill Sutton, "Genetic Engineering in New Zealand," a lecture delivered to the New Zealand Institute of Chemistry at Hamilton, August 24, 1977, p. 13.

4. Jacob Bronowski, *Science and Human Values* (London: Hutchinson, 1961), p. 71.

5. See Jacob Bronowski, "The Disestablishment of Science," in *The Social Impact of Modern Biology,* ed. Watson Fuller (London: Routledge and Kegan Paul, 1971).

6. Compare the views that were expressed by the McCarthy Commission on a proposal concerning the establishment of an independent DSIR. See "The Report of the Royal Commission of Inquiry on the New Zealand State Services," June 1962, paragraphs 128-37. Note also

---

*Compare some proposals concerning committee structures and functions in Chapter 4, section on "A New Statute."

the role of the National Research Advisory Council and the University Grants Committee in channeling government funds to science without direct government intervention.

7. Charles P. Snow, *Science and Government* (London: Oxford University Press, 1961), p. 69.

8. See Peter Steinfels, "Is Science Stoppable," *Commonweal* 8 (1970).

9. Herbert Marcuse, *One Dimensional Man: Studies in the Ideology of Advanced Industrial Society* (Boston: Beacon Press, 1964).

10. See Robert Jungk, *Brighter than a Thousand Suns: A Personal History of the Atomic Scientists* (New York: Harcourt Brace, 1958) for a description of the Manhattan Project and the Einstein (Szilard) letter.

11. However, note that, largely in response to the events of the 1930s and 1940s, an international society of scientists, which is known as the Pugwash Union, was formed in 1957 by Bertrand Russell, with the support of Leo Szilard and Albert Einstein. Its concern is the relationship between science and society. Its meetings are conducted in private and the circulation of most of its publications is restricted. See Joseph Rotblat, *Scientists in the Quest for Peace: A History of the Pugwash Conferences* (Cambridge, Mass.: MIT Press, 1972).

12. See Kenneth J. Aldous, *New Zealand Science Review* 34 (1977): 112. The association has adopted a very limited *Code of Practice for Scientists*, see *New Zealand Science Review* 30 (1972): 33.

13. For example, the U. S. National Science Foundation, which has issued several statements indicating concern with ethical issues, does not employ any experts in this field. It has allocated between $200,000 and $400,000 for ethics-related research out of an annual budget ranging from between $200 million to $400 million. See Willard Gaylin, "Scientific Research and Public Participation," *Hastings Centre Report* 5 (1975): 6.

14. This argument is often used as a justification for the establishment of "royal commissions of inquiry." See, for example, "Nuclear Power Generation in New Zealand," Report of the Royal Commission of Inquiry, April 1978.

15. See the Health and Safety (Genetic Manipulation) Regulations (UK) 1978, No. 752.

16. See "Looking for the Genes That Throw the Cancer Switch," *New Scientist*, August 3, 1978, p. 342.

17. For example, somatostatin and insulin. See also "How Genentech Made Human Insulin," *New Scientist*, September 28, 1978, p 926.

18. "Genetic Engineers Plug Brain Gene into Bacteria," *New Scientist*, November 10, 1977, p. 333; and "US Company takes Genetic Engineering to Market," *New Scientist*, December 8, 1977, p. 619.

19. "Hope for Diabetics," *Evening Post*, September 8, 1978; and also "Genetic Engineers Make Human Insulin," *New Scientist*, September 14, 1978, p. 747.

20. George Wald, "The Case Against Genetic Engineering," *The Sciences*, September/October 1976. Compare E. S. Anderson, "Genetic Engineering — the *Manifest* Hazards," *New Scientist*, July 6, 1978, p. 34.

21. U.S. House of Representatives, Committee on Science and Technology, "Genetic Engineering, Human Genetics and Cell Biology," Supplemental Report II, December 1976, p. 38.

22. Stanley Cohen, "Recombinant DNA: Fact and Fiction," *Science* 195 (February 18, 1977): 654-57.

23. Richard Offinger speaking at the American Association for the Advancement of Science. Extracts from his speech were reported in "Stop Telling People They Are Ignoramuses on DNA," *New Scientist*, March 9, 1978, p. 674.

24. "Intimidation of DNA critics?" *New Scientist*, March 9, 1978.

25. Joshua Lederberg, "DNA Splicing: Will Fear Rob Us of Its Benefits?" *Prism* 3 (November 1975): 33-37.

26. Wald, op. cit.

27. See C. Weldon Jones et al., "Inter-Kingdom Fusion between Human (He La) Cells and Tobacco Hybrid (GGGL) Protoplasts," *Science* 193 (July 30, 1976): 401-03, and "Plant/Animal Hybrids Create a New Era in Biology," *New Scientist,* July 29, 1976, p. 211.

28. Bernard Davis, "Evolution, Epidemiology and Recombinant DNA," *Science* 193 (August 6, 1976): 442. This idea has received some support from experiments conducted by S. Chang and S. N. Cohen. They believe that when the necessary enzymes are present in the bacterium, *E. coli* can recombine pieces of its own plasmid with foreign DNA. See *Proceedings of the National Academy of Sciences* (U.S.) 74 (1977): 4811.

29. See *National Institutes of Health Environmental Impact Statement on NIH Guidelines for Research Involving Recombinant DNA Molecules,* Part I, October 1977, p. 25.

30. On the other hand scientists who attended the Falmouth (Massachusetts) Workshop on Studies for Assessment of Potential Risks Associated with Recombinant DNA Experimentation, June 20-21, 1977, concluded that K-12 is so severely enfeebled that it is not capable of being converted into a pathogen. It has also been argued that the virus SV40, which infects humans, should no longer be used as a vector. See *NIH Environmental Impact Statement,* op. cit., p. 81.

31. "Nuclear Power Generation in New Zealand," op. cit., p. 211.

32. "Genetic Engineering, Human Genetics and Cell Biology," op. cit., p. 40.

33. Eleanor Lawrence, "Porton Lab Will Study Fever Virus," *Nature* 25 (May 15, 1975): 185. Also *Morbidity and Mortality Weekly Report,* Center for Disease Control, Atlanta, U.S. Department of Health, Education and Welfare, Public Health Service, December 23, 1975, pp 378, 383.

34. See Introduction to this book, note 1 and "UK Doctors Confirm Smallpox," *Evening Post,* August 29, 1978, and "Mother Now Victim of Smallpox," *Evening Post,* September 15, 1978.

35. See *Evening Post,* August 9, 1977.

36. Sutton, op. cit.

37. "Crop Genetic Research Aids Farmers," *Dominion,* March 20, 1978.

38. Governments may have an additional interest in using genetic engineering as a means of biological warfare, although that course of action is now prohibited by an international convention. However, open-air biological weapon testing was alarmingly common in the United Kingdom and the United States. Some of the tests have caused irreparable damage. For example, in December 1942 the British government populated Gruinard Island (which is 270 meters off the Scottish mainland) with sheep and cattle and sprayed it with anthrax-infected liquid. The sheep and cattle died and their carcases were buried in caves that were then sealed. No attempt will be made to rid the deserted island of the anthrax spores that still exist there — a government spokesman has stated that the task would be too "immense and costly." See "Lethal Anthrax Spores Remain," *Evening Post,* May 9, 1978. The U. S. Army has admitted that it conducted 239 open-air germ warfare tests in five cities (Washington, D.C., New York, San Francisco, Key West, and Panama City) in the period between 1959 and 1969. These tests included the release of supposedly harmless "bugs" in Washington's Greyhound bus terminal and international airport. In the 1950 San Francisco experiments the bacterium *Seriatia marcescens* was used. Medical researchers suspect that it may have caused ten cases of pneumonia in the San Francisco area and one death. See "Army Conducted 239 Secret, Open-Air Germ Warfare Tests," *Washington Post,* 1978. Similarly, there is a possibility that terrorists will attempt to use these techniques. Lederberg, op. cit., has suggested that recombinant DNA research (which involves the most complex of the genetic engineering techniques) "is simple enough to be applied in any laboratory which can handle pure bacterial cultures."

39. This issue lies at the heart of the debate surrounding recent claims that a human being has been cloned. See David M. Rorvik, *In His Image: the Cloning of a Man* (Philadelphia: Lippincott, 1978).

40. Clive S. Lewis, *The Abolition of Man; or Reflections on Education with Special Reference to the Teaching of English in the Upper Forms of School* (New York: Collier Books, 1962), pp. 70-71.

41. For example, Ehrlich suggests that it would be unfortunate if blanket "class" bans on organisms like *Salmonella* were to prevent the further development of the Ames' test, which provides an extremely important means of detecting environmental carcinogens. See Paul Ehrlich, "An Open Letter to the Board of Directors of Friends of the Earth," *Coevolutionary Quarterly*, Spring 1978.

42. See Chapter 3, first section, for a review of the *National Institutes of Health [U.S.] Guidelines for Research Involving Recombinant DNA Molecules. Federal Register* 41 (July 7, 1976): 27902.

43. Compare the Health and Safety (Genetic Manipulation) Regulations 1978 (U.K.), No. 752.

3

TOWARDS NATIONAL CONTROL

Many countries have already taken steps toward controlling genetic engineering research. In the United Kingdom this control takes the form of legislation. In New Zealand nonmandatory guidelines have been formulated, although the possibility of legislation is currently under consideration. The following analysis of the diverse methods of control that have been adopted should be of some use in assisting to determine which types of decision-making and enforcement procedures are most likely to be effective.

THE UNITED STATES

The first attempts to control genetic engineering research were made in the United States. The work that was being done in this field was brought to the attention of the U.S. government in 1971 when James Watson addressed the House Science and Astronautics Committee on the subject of genetic engineering.[1] Motivated in part by Watson's remarks, the committee directed the preparation of a report on this research. Three reports followed.[2] They described the need for a continuing awareness of the risks involved in genetic engineering experiments and highlighted the rapid rate at which knowledge in this field was expanding. No public indication of concern about genetic engineering techniques was forthcoming from the scientific community until June 1973 when the Gordon Conference on

Nucleic Acids was convened in New Hampshire. Several of the speakers warned of the potential dangers involved in recombinant DNA research and an open letter was sent to the president of the National Academy of Sciences.[3] The letter called for the establishment of a study group to consider recombinant DNA techniques and to recommend guidelines. The National Academy formed a committee and as a result another letter was published, this time by the members of the committee.[4] Their controversial letter contained three main proposals. The first was to the effect that scientists throughout the world should voluntarily defer certain specified experiments until formal guidelines were drawn up to control the research. The second was that the director of the National Institutes of Health (NIH) should establish an advisory committee to investigate the dangers involved in recombinant DNA research and to devise guidelines for workers in this field. The third involved a call for an international meeting of scientists at which these issues could be discussed.

All of these proposals were acted upon and as far as can be ascertained it appears that there is universal compliance with the voluntary ban. On October 7, 1974, an NIH recombinant DNA Molecule Advisory Committee was set up to review that type of research and to formulate guidelines. In February 1975 scientists from 15 countries met at Asilomar in California to discuss recombinant DNA techniques. Of the 155 people who were present, 83 represented research, governmental, and industrial institutions in the United States, 51 were foreign scientists, 16 were representatives of the press, 4 were lawyers, and 1 was a lay person.[5] Attendance was by invitation only and the papers and discussions were not published, although a summary statement was issued.[6]

It is unfortunate that the scientists involved did not attempt to attract a wider audience. The agenda specifically referred to ethical as well as to purely scientific matters, and while it is encouraging to note that four lawyers were invited,* it is difficult to understand why the views of specialists in other fields were not sought.† Nevertheless it is clear that the conference was enormously influential. In fact the recommendations embodied in the summary statement formed the basis of a nonmandatory set of guidelines that were published by the NIH in the *Federal Register* on July 7, 1976.

---

*One of the lawyers, Daniel Singer, of the Institute of Society, Ethics and Life Sciences, Hastings-on-Hudson, New York, is a specialist in the field of morality and ethics.

†The conference can be compared with a series of forums and committee meetings that were open to the public between November 1975 and May 1976 at the University of Michigan at Ann Arbor. Of three university committees that were charged with deciding whether a P3 Laboratory should be constructed at the university for the purposes of recombinant DNA research, one committee was composed of lawyers, social workers, philosophers, theologians, and scientists drawn from fields other than molecular biology. The university eventually voted in favor of proceeding with the construction of the facility.

The scientific community had not been able to agree upon the need for guidelines. Even the views of those who suggested that there was a need for control were polarized. Some favored biological containment, some favored physical containment, and others favored a combination of methods. Similarly, there was no identifiable consensus concerning which types of experiments, if any, should be prohibited. The guidelines represented a compromise between these groups. Some recombinant DNA experiments are expressly banned. For example, certain types of experiments involving more than ten liters of culture are not to be carried out with recombinant DNA molecules. All the prohibitions are, however, qualified by the inclusion of a proviso that gives the director of the NIH a discretion to approve these experiments after considering scientific and societal risks and benefits.

As regards permissible experiments, four categories of physical containment laboratory are defined. P1 facilities are of the type that are commonly used for microbiological work involving little or no risk.[7] P2 laboratories incorporate special safety devices such as autoclaves and safety cabinets. Subject to more restricted access to the laboratory, the level of confinement is, however, similar to P1 stringency. A "moderate" level of safety is provided by P3 facilities, which must incorporate special engineering and design features and physical containment facilities such as air locks and ventilation systems.[8] P4 facilities provide the highest degree of containment.[9] The laboratory is either in a separate building or completely isolated from other parts of the main building. Operational procedures are set out in full in the guidelines. Considerable emphasis is also placed on biological containment and three levels of confinement that relate principally to host-vector systems are described. Lists of permissible recombinant DNA experiments are then linked with appropriate containment categories and laboratory practices.[10]

The decision-making and approval process is of particular interest. The guidelines envisage a tiered system of control. In the first instance institutions involved in recombinant DNA research are required to establish "biohazard committees." These committees serve to encourage compliance with the guidelines at the local level and to provide a link with the NIH. That link facilitates the application of the suggested uniform national standard that is embodied in the guidelines. One of the principal roles of the local committees is to act as a source of advice and information for the researcher. They are also responsible for certifying and annually recertifying the safety and suitability of facilities and procedures that are used in the institution. In addition, they have some responsibility for training laboratory staff. The minutes of their meetings are available for inspection by the public but members of the local community do not sit on the committees, which are entirely composed of experts in technical fields such as biology, safety, and engineering. Nevertheless, there is nothing to preclude the members of the committee from consulting with nonscientific advisors if they wish to do so. The principal researcher is required to seek specific NIH approval for each recombinant DNA experiment. The application can then be considered by the NIH Recombinant DNA Molecule Program Advisory Committee. With the exception of an expert in government and public affairs, who sits on the Recombinant DNA Advisory Committee, both these bodies have a purely scientific membership that consists principally, although not entirely, of scientists who represent disciplines that are actively engaged in recombinant DNA research. The decision as to whether or not to approve an application generally rests with the Recombinant DNA Advisory Committee. However, if that committee is uncertain it can refer the matter to the Advisory Committee of the director of the NIH. It is felt that this committee, which is at present of uncertain composition, will "serve to provide the broader public policy perspectives."[11] Presumably, the final decision lies with the director of the NIH in view of the fact that a right of appeal is not mentioned.

The NIH has placed particular emphasis on the responsibilities of individual researchers. For example, they are required to assess the hazards involved in their experiments, to plan procedures that would minimize risks, to seek NIH approval, and to inform the NIH in writing if even a minor accident occurs within the confines of the laboratory. Since the guidelines are not enforced by law, how meaningful are these responsibilities? Laboratories that are funded by the NIH are effectively bound by the guidelines. In fact their responsibility is expressly linked with NIH sponsorship.[12] Funded investigators were asked to give an assurance that they would comply with the guidelines and although no direct threat of withdrawal of funds was made, the implication was clear enough.

Largely as a result of the work of an interagency committee, several government departments that are not funded by the NIH have voluntarily agreed to observe the guidelines. They include the National Science Foundation, the Energy Research and Development Administration, the Department of Defense, and the Department of Health, Education and Welfare. Nevertheless, several important government agencies, such as the Department of Agriculture, have not indicated that they will comply with the guidelines. In introducing the guidelines the director of the NIH expressed the hope that industry would adopt them as their code of practice.* Obviously, corporate compliance would be voluntary as the funding control that is influential in the governmental sphere does not usually operate in

---

*The director held a private meeting with the representatives of 20 U.S. corporations on June 2, 1976.

relation to companies. In view of the fact that not all government departments doing research in this area have indicated that they will adhere to the guidelines, the director may have been a little optimistic. Corporate involvement in this research is already considerable in the United States.[13] In March 1977 genetic engineering experiments were being conducted by General Electric, which had already applied for a patent on one of its microorganisms, and also by Miles Laboratories, Eli Lilly and Company, Smith, Kline and Company, Hoffman-La Roche, Genentech, the Upjohn Company, Merck Sharpe and Dohme, Pfizer Incorporated, and Abbott Laboratories. Several other companies that specialize in the manufacture of chemicals, agricultural products, and pharmaceuticals are currently investigating the possibility of entering this field. They include Cetus, CIBA-Geigy, du Pont, Dow, W. R. Grace, Monsanto, French, Wyeth, and Searle Laboratories.

In September 1976 C. J. Stetler, the president of the Pharmaceutical Manufacturers Association, indicated that the drug companies endorsed the "spirit and intent" of the guidelines and would endeavor to follow them.[14] Such a response is obviously in the interests of the companies in the sense that a less than positive reaction to the guidelines is likely to result in increased awareness of the need to regulate corporate activities by legislation. In spite of that factor, General Electric and the Manufacturing Chemists Association did not accept Senator Kennedy's invitations to discuss the guidelines with his Subcommittee on Health.[15]

The problem of uncontrolled corporate research becomes more obvious as both the scale of the experiments and the number of companies involved increases. For instance, the prohibition on experiments including more than ten liters of culture, which is likely to be ignored by companies, will be most relevant in the industrial context. It is also probable that it is the profit-making organization that will be least inclined to use manpower and money to improve containment facilities and minimize risks.

Companies are not likely to be totally insensitive to public opinion, although even the NIH can be criticized for failing to take adequate account of community views. When introducing the guidelines the director stated that "it must be clearly understood by the reader that the material that follows is not proposed rule making in the technical sense, but is a document on which early public comment and participation is invited."[16] Unfortunately, this invitation was a little belated. Apart from a public meeting that had been held on a set of draft guidelines, concerted attempts to involve the public do not appear to have been made. In fact the NIH has more than a moral duty to consider public opinion. Under the National Environmental Policy Act 1969[17] it has a legal obligation to do so. That duty was breached when the NIH failed to prepare an environmental impact statement prior to the publication of the guidelines. The National

Environmental Policy Act (NEPA) commences with a statement of intent. Two of the four stated purposes of the act are: "To declare a national policy which will encourage productive and enjoyable harmony between man and his environment" and "To promote efforts which will prevent or eliminate damage to the environment and biosphere and stimulate the health and welfare of man . . . ." In addition, section 101 refers to the "interrelations of all components of the natural environment" and "the overall welfare and development of man."

The environmental impact statement is the principal means by which those broad policy statements are transformed into actions and procedures that are likely to result in environmental protection. Accordingly, in section 102(2) (c), Congress requires that all agencies of the federal government are to include a detailed statement by the responsible official, in every recommendation or report on proposals for legislation and other major federal actions "significantly affecting the quality of the human environment." Section 101(c) recognizes that each person should enjoy a healthy environment and that each person has a responsibility to contribute to the preservation and enhancement of the environment. Thus the impact statement is to be made available to the public, by means of publication in the *Federal Register*.[18] Public participation is not, however, confined to the presentation of comments on a completed environmental impact statement. The public must be notified when a draft statement is about to be prepared and no administrative action can be taken until 90 days after the draft has been circulated for comment and until 30 days after the final text of the statement has been made available to the public. The federal agency that is responsible for a project can determine whether it will hold a public hearing on a draft impact statement; if it does it must issue the draft at least 15 days prior to the date of the hearing. The NIH did not attempt to deny that the Recombinant DNA Molecule Guidelines fell within the ambit of the act and it published both the draft and a final environmental impact statement on the guidelines,[19] but it did not hold a public hearing and, even more importantly, it did not publish either the draft or the final statement before it issued the guidelines. An individual or an environmental agency can take this matter to court. [20] In the past the courts have found their jurisdiction in section 10 of the Administrative Procedure Act,[21] which subjects "each authority of the government of the United States" to judicial review except where there is a statutory prohibition on review.[22] They have also been active in enforcing the preparation of impact statements "as early as possible and in all cases prior to Agency decision."[23]

Although the main function of the courts is to enforce compliance with the procedural requirements of the National Environmental Policy Act, they can also exert some influence over the substantive content of the guidelines. For example, if the impact statement is challenged on the basis that it is ill-considered and hastily prepared, then the court can direct the preparation of a new report that takes full account of specified environmental factors.

Despite the fact that the late publication of the statement deprived the public of an opportunity to participate in formulating the guidelines, it was felt that the environmental impact statement would not be challenged, possibly because it represented a pedestrian but reasonable attempt to identify and evaluate the issues involved in recombinant DNA research.[24] The need for control was accepted and various methods by which it could be achieved were considered. These included an NIH prohibition on the funding of all recombinant DNA experiments, the formulation of a different set of guidelines, and the possibility of abandoning the guidelines in favor of careful NIH consideration of each proposed project before funding. As might have been expected, the statement supports the guidelines. The closest that it comes to heresy is in failing to defend criticisms that it notes were made of the decision not to prohibit "shotgun" experiments and experiments that involve the use of DNA from tumor-producing viruses. Hopefully this tacit disapproval will not go unnoticed when the NIH revises the guidelines.

On May 9, 1977, the Friends of the Earth filed a lawsuit in the federal district court of the Southern District of New York.[25] The action for declaratory and injunctive relief consists of seven claims. The three principal claims are that the NIH failed to comply with NEPA by not preparing an environmental impact statement prior to funding the recombinant DNA research program and promulgating the NIH guidelines; that in issuing the guidelines the NIH failed to comply with procedural notice and comment provisions in the Administrative Procedure Act; and that the Recombinant DNA Advisory Committee, which drafted the NIH guidelines, did not have a balanced membership "in terms of the points of view represented and the functions to be performed" as required by the Federal Advisory Committee Act.[26] The case, which has not yet been decided, was not intended to have the effect of halting recombinant DNA research per se. The organization's aim was simply to place decision making on recombinant DNA activities on a solid base and to ensure that there are opportunities for effective public participation.[27]

The plaintiff in *Mack* v. *Califano* had a different objective.[28] He was a resident of Frederick, Maryland, who on May 31, 1977, filed a motion for a temporary restraining order and a preliminary injunction to prevent the defendants from undertaking certain P4 recombinant DNA experiments at the federal research center at Fort Detrick.* On July 18, 1977, it was agreed that the project would be discontinued until after the publication of the

---

*The project involved experiments with polyoma, a virus known to cause cancer in mice. The vector to be used was a derivative of *E. coli*, K-12.

final environmental impact statement on the NIH guidelines. Subsequent to the publication of the statement the plaintiff proceeded with his motion for a preliminary injunction on the ground that the statement did not comply with the requirements of the National Environmental Policy Act. The court disagreed and refused to grant the injunction. Smith, J. was of the opinion that the NIH had carefully considered the potential risks of the experiment and had taken the precautions that were necessary to prevent harm to the environment. He was clearly prepared to accept the environmental impact statement on the NIH guidelines as evidence of the appraisal of the risks involved in the proposed Fort Detrick experiment and it is especially interesting to note that in the penultimate paragraph of his judgment, Smith, J. emphasized that "the experiment is designed to provide important and needed information on the possibilities of recombinant DNA technology. Important scientific information relative to the possibilities of this technology would be delayed if a preliminary injunction were granted."[29] Although it may be possible to limit the case to its own facts, the decision in *Mack* is significant in the sense that it appears to represent a refusal to apply NEPA's provisions to particular experiments. All federal agencies are required to prepare impact statements on "major federal actions significantly affecting the quality of the human environment." Accordingly, the decision of the court can be justified only on the basis that the action was not sufficiently "major" or that it did not have the potential to significantly affect the quality of the human environment.

In the past, two approaches to the question of whether or not a project can be classified as a major one have been adopted. The first and more common view is that the determining factors in the assessment are the cost of the project, the amount of planning that precedes it, and the time required to complete the project, rather than the environmental impact of the action.[30] The other approach involves an assumption that the magnitude of a federal action cannot be separated from the significance of its effect on the environment. Thus, if an action has a significant impact on the environment it is necessarily a major action. The latter analysis is unfortunate in the sense that the preparation of a statement is mandatory only when the action is one that significantly affects the environment.[31] If the first test is applied to the facts of the *Mack* case it would appear that the criteria of cost, planning, and completion time are not satisfied. However, in real terms, expensive and ill-considered projects are probably more likely to result in environmental damage than their well-planned counterparts. Hence it is arguable that there should be only one relevant test, namely, that which relates to a potentially significant impact on the environment.

Proposed revised guidelines in which public comment was invited were published in September 1977 and a public meeting to discuss these guidelines was convened by the NIH on December 15, 1977.[32] The proposed revisions represented a relaxation of the requirements in the guidelines and have been criticized because the data on which they were founded were not published.[33] In the meantime, the Recombinant DNA Advisory Committee has drafted another set of proposed revisions that may even further weaken the guidelines.[34] The suggested amendments differ quite markedly from those that were publicly discussed in December. For example, the NIH has delegated the initial authority for the approval of experiments to the local biohazard committees. This diminishes the possibility of the application of a uniform national standard. In introducing the first set of guidelines, the director of the NIH had commented that he was reluctant to let local biohazard committees approve experiments because such a practice would be likely to threaten uniformity of standard.[35]

Public comment on regulatory proposals is not the only possible avenue of public participation. The public can in fact play a more direct role in controlling genetic engineering as, for example, through local body control. On June 23, 1976, the same day that the NIH guidelines were released and one week before they were published in the *Federal Register*, the City Council of Cambridge, Massachusetts, began a series of public hearings on recombinant DNA research. This research is conducted at Harvard University and the Massachusetts Institute of Technology, but it was the controversial Harvard Laboratory that prompted the hearings. The first meeting, which was attended by more than 500 people, was called by the mayor of Cambridge, after Harvard announced plans to rebuild one of its laboratories to P3 containment standards.[36] Both proponents and opponents of recombinant DNA research presented their views at the meeting. A second hearing was held on July 7, and on that occasion the nine-member City Council voted by five to three, with one abstention, to declare a three-month "good faith" moratorium on recombinant DNA research that would require the use of P3 or P4 facilities.* In practical terms, the council did not have the power to vote for anything more stringent than a good faith moratorium, because the only legal authority to ban research was vested in the city's health commissioner, who could declare the work to be a health hazard. This was a matter of some embarrassment for the mayor because the city had not had a health commissioner for the previous 19 months.

In addition to the proposed moratorium, the council voted to establish a committee to be known as the Cambridge Laboratory Experimentation Review Board. It would investigate recombinant DNA research and make

*MIT already had a P3 laboratory but was not yet using it because the voluntary Asilomar moratorium had ceased with the publication of the NIH guidelines only two weeks before the Cambridge City Council made its decision.

recommendations to the council concerning the work being conducted at Harvard and MIT. The board, which was composed of city residents, presented its conclusions to the Cambridge City Council on January 5, 1977. It concluded unanimously that experiments of the type that require the use of P3 facilities could be carried out in Cambridge only if certain requirements, additional to those specified in the NIH guidelines, were fulfilled.[37] For instance, it was proposed that local biohazard committees must contain at least one member who is not connected with the institution that is conducting the research. It was also stated that biologically disabled microorganisms should be used in all P3-level experiments and that every possible attempt should be made to monitor the escape of host organisms. The board also suggested the establishment of a Cambridge Biohazards Committee for the purpose of overseeing all recombinant DNA research conducted in Cambridge. Its members were to be city residents who had no affiliations with the universities. Its functions would be to review research proposals for the purpose of ascertaining whether the NIH guidelines would be complied with, to maintain links with the Havard and MIT Biohazard Committees, and to establish procedures that would enable members of the relevant research institutes to inform the Cambridge Biohazard Committee of any breaches of either the NIH guidelines or the additional Review Board recommendations.

The Review Board did not limit itself to making proposals for the City of Cambridge. It requested Congress to convert the NIH guidelines into legally binding regulations that would apply to recombinant DNA research throughout the United States. The board also suggested that a national registry of laboratory workers should be established in order to aid long-term epidemiological studies. On February 7, 1977, the City Council voted by six votes to three to lift its ban on P3-level research and to incorporate the board's recommendations into the city's ordinances.

It has been argued that the unprecedented decision-making role that was adopted by the people of Massachusetts was paralleled and indeed prompted by a certain amount of animosity between citizen and researcher in Cambridge. There appears to be little doubt that the mayor of Cambridge has, in the past, made political capital out of threats of depriving the university of its tax exemptions and its land.[38] However, the incident cannot simply be dismissed in that way. The Review Board Recommendations received qualified praise from scientists who represented opposing factions in the control controversy and this success has not gone unnoticed in San Diego and New York, where methods of controlling recombinant DNA research are currently under consideration.

The events in Cambridge provide a concrete example of the practicality of public participation in decision making.* Perhaps the citizens' right to

---

*Local ordinances have also been passed in Princeton, New Jersey.

decide can be exercised in a particularly effective manner when scientific opinion is divided. But while the Cambridge community succeeded in entering into the decision-making arena, their efforts, like the NIH guidelines, lack the force of the legal sanction. They rely on the good faith and the voluntary compliance of the researchers. These factors have not proved to be sufficiently strong means of control. Both the NIH guidelines and the citizens' recommendations have already been disregarded by one of the Harvard researchers.

As a result of a request for information by a member of the Environmental Defense Fund under the Freedom of Information Act of 1966, the NIH instructed Charles Thomas of the Harvard laboratory to stop conducting recombinant DNA research pending an inquiry.[39] A three-man NIH team was sent to Harvard, where it became apparent that approval for the experiments had not been sought, either from the Harvard Biosafety Committee or the NIH. There was no evidence of the so-called Memorandum of Understanding and Agreement (MUA), which was required by the guidelines, and the relevant research grant application did not mention recombinant DNA research. The NIH team is still investigating the situation and the possible use of P2 facilities for experiments requiring a P3 level of containment. Although the NIH responded promptly in the first instance, no one now seems to know what to do next. One NIH official has stated: "If we find out Thomas has been doing recombinant DNA research without an MUA, then some appropriate action is going to have to be taken, but what that is, I don't know."

The Harvard incident has not been the only recorded case of noncompliance with the guidelines. The first instance of a breach had occurred in mid-1977 at the University of California in San Francisco.[40] Researchers in the Departments of Biochemistry and Biophysics conducted an experiment with a vector that had not been certified by the director of the NIH. The incident gave rise to some concern about the way in which scientists at the university viewed the guidelines. For instance, Hopson has claimed that "half of the researchers here follow the guidelines fastidiously, others seem to care little . . . ."[41] She also noted that "among the young graduate students and post-doctorates it seemed almost chic not to know the NIH rules."

If there is a tendency to flout the guidelines in some laboratories, it is likely that in those instances the NIH will soon become aware of breaches. In the particular case of the experiment that caused the controversy, it appears that information concerning a possible breach came from within the university itself. In this context, it must be remembered that in December 1977 the NIH was sponsoring 250 recombinant DNA research programs in 110 different institutes. Under those conditions, it is clear that serious attempts to ensure that the guidelines are complied with cannot be made without the help of a sizable inspectorate.

Although the researchers at the University of California abandoned their experiment before it was completed, it seems likely that the information that they gained from their work with the uncertified vector was at least partly responsible for an important development they pioneered within three weeks of the commencement of the same experiment with a newly certified vector. When a breakthrough is imminent, there is obviously an enormous incentive to experiment with uncertified vectors. The San Francisco team was convinced that the Harvard Laboratory was also on the verge of a breakthrough in this field.* In this type of situation there is a strong need for an efficient administration operating an efficient approval system. If guidelines or recommendations are to command the respect of the scientific community, they must take cognizance of the scientific method. On the other hand it is almost inevitable that the procedural requirements that are set out in guidelines will have some stultifying effect on research in a rapidly developing field. That effect is aggravated by the competitive nature of the work concerned.

One of the more important points illustrated by the experiment was the ineffectiveness of both the Biosafety Committee and the NIH response. The chairman of the university's Biohazard Committee investigated the incident but his comments consisted mainly of criticisms of the slowness of NIH certification and notification procedures. The swift and positive NIH intervention in the Harvard case was presumably attributable to sensitivity to the criticisms that had been directed at the NIH after the San Francisco incident. In spite of the improved response, it is clear that neither the recommedations of the citizens of Massachusetts nor the more formalistic NIH guidelines represent successful attempts to control science and to make scientists accountable to the public.

Many of those who have commented on the recombinant DNA controversy have expressed the opinion that the sanction that is lacking will have to be provided by legislation. For example, on April 26, 1977, the National Academy of Sciences (NAS) resolved that "the NIH guidelines are the result of careful deliberation and we favor their simple conversion into a uniform national set of regulations." Initially that view was not shared by the director of the NIH. Nine months prior to the NAS resolution, he had stated that the flexibility of the guidelines must be preserved and that he shared the concern of those commentators who had suggested that the implementation of the guidelines as regulations would introduce unnecessary rigidity.[42] The rationale for this was that legislative amendments would not be likely to keep pace with scientific developments.

---

*Harvard's research was not, in fact, as advanced as had been imagined. Compare the adverse effects of the secrecy surrounding atomic research in the 1930s.

In October 1976 the secretary of the Department of Health, Education and Welfare (HEW) established an interagency committee on recombinant DNA research under the chairmanship of the director of the NIH. The committee's terms of reference authorized it to review the nature and scope of federal and private sector recombinant DNA research and to recommend appropriate legislation or executive action. The series of meetings that followed appear to have had a profound influence on the director. At the public meeting that was called in December 1977 to discuss the proposed NIH revised guidelines, a changed attitude was evident. He said, "My own belief is that it would be to the maximum advantage of the country for a very simple legislative package to be passed extending the existing guidelines to everyone."

That was not the unanimous opinion of those who attended the public meeting. Some disagreed both with attempts to incorporate the guidelines into an act and with the idea of passing new legislation to control this research. Congressman Ottinger, Peter Hutt (former general counsel for the Environmental Defense Fund), the National Resources Defense Council, and the Food and Drug Administration have been vocal supporters of control by means of existing legislation. They advocate the use of section 361 of the Public Health Service Act,[43] which gives the secretary of HEW extensive powers to control communicable diseases.[44] In this way, they argue, the standards embodied in the NIH guidelines could be implemented as regulations and enforced against organizations that are not associated with the NIH.

The interagency committee delegated the task of reviewing the possibility of control, by means of existing legislation, to a subcommittee, but before the investigations began the Interagency Committee shaped the form of the subcommittee's conclusions by prescribing that control under the provisions of existing legislation would not be acceptable to the government. It would be necessary to ensure compliance with four specified minimum requirements:

Review of such research by an institutional Biohazards Committee before it is undertaken

Compliance with physical and biological containment standards and prohibitions in the NIH guidelines

Registration of such research with a national registry at the time the research is undertaken (subject to appropriate safeguards to protect proprietary interests)

Enforcement of the above requirements through monitoring, inspections, and sanctions.[45]

The subcommittee discussed sections 361 and 353 of the Public Health

Service Act. They disagreed with those who had suggested that the act could, without amendment, be used to regulate recombinant DNA research. In order to apply section 361 to all recombinant DNA research, there would have to be a reasonable basis for concluding that the products of all recombinant DNA research can cause communicable human disease. In addition, it was argued that section 361 does not apply to plants, animals, or the environment. Section 353 of the act, which empowers the Center for Disease Control to license and control the operation of clinical laboratories, was not considered to be applicable to research laboratories.

The Occupational Safety and Health Act 1970 was also reviewed.[46] The subcommittee noted that the act gives the Occupational Safety and Health Administration (OSHA) broad powers to require employers to provide a safe workplace for their employees. However, OSHA standards do not extend to the self-employed, nor do they bind individual states unless they are voluntarily adopted — in fact, 26 states are not subject to the OSHA standards. The act obviously does not meet the Interagency Committee's criteria. Similarly the Hazardous Materials Transportation Act[47] was thought to be too limited to be of use. It regulates the interstate shipment of "hazardous materials," including biological products. The act is administered by the Department of Transportation and is primarily concerned with packaging, labeling, and shipping requirements.

The Toxic Substances Control Act[48] is arguably the most suitable legislation for the purpose of controlling genetic engineering. It is administered by the Environmental Protection Agency, which, by virtue of section 6 of the act, has the responsibility for controlling the "manufacture, processing, distribution in commerce, use, or disposal of a chemical substance which will represent an unreasonable risk of injury to health or the environment." A "chemical substance" is defined in section 3 of the act and includes any organic or inorganic substance of a particular molecular identity. This potentially all-embracing definition clearly covers the materials used in recombinant DNA research. On the other hand, section 5 explicitly exempts the registration of chemical substances used in small quantities for the purposes of scientific experimentation or analysis. In relation to the wider question of the registration of researchers engaged in recombinant DNA research, the Environmental Protection Agency has stated that it does not have the authority to compile such a register. That is a fatal deficiency in terms of the minimum requirements set out by the Interagency Committee.

The subcommittee also considered the possibility of regulation, by the Environmental Protection Agency, under the Clean Air Act, the Federal Water Pollution Control Act, and the Resource Conservation and Recovery Act.[49] Under section 112 of the Clean Air Act, the Environmental Protection Agency can list hazardous air pollutants and set emission standards.

Under section 307 of the Federal Water Pollution Act, toxic effluent standards can be set for toxic pollutants. However, the agency has suggested that it does not have sufficient information about the hazards of DNA processes to enable it to formulate such standards.[50] In addition it questions its suitability for the difficult task of monitoring discharges and enforcing standards. Similarly, the regulatory powers of the Department of Agriculture are of little assistance since they do not extend to humans. The possibility of enforcement by the Food and Drug Administration (FDA) was also discounted on the basis that the commercial use of these techniques was not sufficiently prevalent to warrant FDA intervention. Under these circumstances the subcommittee quite properly concluded that the minimum standards that were set out by the Interagency Committee could not be enforced under existing legislation and the Interagency Committee accepted that conclusion. It was clear that no existing statute was suitable for the purpose of control and equally clear that the solution could not be found in the implementation of a combination of different provisions in distinct acts. The adoption of the latter alternative would simply be likely to result in a lack of coordination and an unwieldy administration. Accordingly, the need for the introduction of a single statute for the purpose of controlling recombinant DNA research has recently been recognized by the federal government and two state governments.

On March 15, 1977, the Interagency Committee presented its report to Joseph Califano, then secretary of HEW. He immediately acknowledged the need for a new statute that would regulate recombinant DNA research in the United States. By this time six different recombinant DNA bills had been introduced — one in New York, one in California, and the rest at the federal level. Ten more were to follow in the next ten months. The first of these bills was introduced into the U. S. Senate on February 4, 1977. The Recombinant DNA Research Bill is "to provide for guidelines and strict liability in the development of research related to recombinant DNA."[51] This statement of purpose is followed, in clause 2, by a list of vague congressional "findings" concerning the potential risks and benefits of recombinant DNA research in terms of the need for legislation to protect the public interest.*

The duty to promulgate guidelines under the bill would be placed upon the secretary of HEW, who would have to complete his task within 90 days of enactment. These guidelines would "apply to all research involving recombinant DNA that is in or affects commerce, or that is carried on in any way subject to the jurisdiction of the United States" (clause 5).

---

*Clause 3 defines recombinant DNA as "molecules that consist of different segments of deoxyribonucleic acid which have been joined together in cell-free systems to infect and replicate in some host cell either autonomously or on an integrated part of the host genome."

Clause 7 of the bill provides that persons carrying out recombinant DNA research will be liable, without fault, for all injury to persons or property that is caused by their research. This can simply be regarded as a statutory recognition of the *Rylands* v. *Fletcher* doctrine in the sense that those who engage in hazardous activities are made to bear the loss that results from them. Clauses 8 to 11 of the bill deal with licensing. Clause 8 provides that "no person shall solicit or accept, directly or indirectly, any specimen for research which shall involve recombinant DNA or conduct such research unless there is in effect a licence for such research issued by the secretary." The clause, when read with the penalty provisions described in clause 14, is obviously designed to act as a deterrent, not only to researchers who might have an interest in proceeding without a license, but also to the organizations that employ such researchers.

The authority to issue the licenses would be vested in the secretary of HEW, who has a discretion to issue the licence subject to such terms and conditions as he thinks fit. It is also worth noting that the secretary could not issue a licence unless the information that accompanies the application is sufficient to satisfy the secretary that the guidelines that have been promulgated under the act will be complied with. The licence could be reviewed by the secretary at any time, under clause 11, and revoked, suspended, or limited if, *after* giving reasonable notice and opportunity for a hearing to the licensee, the secretary "finds" that the licensee

> has been guilty of misrepresentation in obtaining a licence
> has engaged or attempted to engage or represented himself as entitled to perform any research or procedure or category of procedures not authorized by the licence
> has failed to comply with guidelines with respect to research facilities or personnel prescribed by the secretary pursuant to this act
> has failed to comply with reasonable requests of the secretary for any information or materials the secretary deems necessary to determine continued eligibility for its licence or continued compliance with the secretary's guidelines
> has refused a request from the secretary or any federal officer or employee designated by the secretary for permission to inspect the research facility and its operations and pertinent records at any reasonable time, or
> has violated or aided and abetted any violation of any provisions of this Section or any guideline promulgated hereunder.

The nature of the hearing that must be granted by the secretary is not specified and the bill does not provide for any right of appeal from his decisions.

Clause 12 is potentially a powerful, if vague, enforcing provision. If the secretary has reason to believe that a continuation of any activity by a research facility licensed under the act would constitute "a significant hazard to the public health," the clause empowers the attorney general, at the request of the secretary, to bring an action in the name of the United States in the court of the district in which the facility is situated. "Upon a proper showing" a temporary injunction or restraining order against continuation can be issued pending the grant of a final order.

Guidance as to the meaning of a significant hazard to the public health is not given, and accordingly the court could have a very important role in determining the effectiveness of the legislation. The final restraining order simply amounts to a judicial prohibition of certain types of recombinant DNA research. While the courts would be unlikely to go beyond the prohibitions and controls specified in the guidelines, it is clear that they will have to grapple with complex scientific evidence. However, in the past, the U.S. courts have demonstrated a willingness and an ability to deal with similar issues, particularly in the context of the National Environmental Policy Act 1969.

The bill also provides for the establishment of an inspectorate. The inspectors are given extensive powers to enter onto premises and to investigate the conduct of the research. But what sanctions are available if a breach is discovered? It is arguable that the bill's penalty clause is quite inadequate for dealing with companies. Any person who *willfully* violates any provision of the act or any guideline made under the authority of the act can be convicted and sentenced to a maximum of one year's imprisonment or a maximum fine of $10,000 or both (see clause 14[a]). In addition the courts may order that any "person or entity" convicted be deprived of federal funds for a period of up to five years. The penalties have the potential to cripple individual researchers, but defaulting companies would remain relatively unscathed. The possibility of a $10,000 fine is not likely to represent an effective deterrent, particularly in a race for a multimillion dollar invention. Finally, clause 14(b) protects employees who assist in investigations or who are responsible for the commencement of proceedings under the act. All these provisions should help to ensure effective policing of the act and its guidelines.

The Commission on Genetic Research and Engineering Bill was introduced into the House of Representatives by Congressman Solarz on March 1, 1977.[52] As its title implies, the purpose of the bill is to establish a commission to investigate genetic engineering research. The commission would consist of 17 members, including two officials from the executive branch of the federal government, two senators from different political parties, two members of the House of Representatives with different political affiliations, and 11 members who are not employees or officials of the federal government and "who, because of their knowledge, expertise, diversity of experience, and distinguished service to their professions, are particularly qualified for service on the commission" (see clauses 3[b] and 3[c]). The 11 would include representatives of state and local governments, members of the academic community concerned with genetic research, and citizens' groups with a special interest in consumer protection. Thus there is a fairly full recognition of the fact that nonscientists have an important role to play in this area even if their inclusion is a matter of political expediency. The commission can, of course, call on the services and advice of experts in a consultative capacity. It also has a duty to publish an annual report to Congress summarizing its activities, findings, and any recommendations that it has for legislation.*

The bill serves as an indication that at least some members of Congress feel that S. 621 was premature and that more information on genetic techniques is required before legislation is passed.† The aim of the bill is to establish an inquiry procedure in order to ascertain the risks and benefits that are involved in the research by encouraging public as well as specialist participation in the fact-finding process. Such an inquiry should provide a firm basis for decisions on legislation.

On March 1, 1977, New York became the first state to introduce a bill for the purpose of controlling recombinant DNA research. The bill took the form of an amendment to a preexisting Public Health Law.[53] It is similar to S.621 but it also proposed the establishment of an advisory committee that would be of similar composition to the Commission envisaged in the Genetic Research and Engineering Bill. The power to make regulations under the act is vested in the hands of the State Health Commissioner whose department would act as an inspectorate. Although the bill confers very wide powers on the commissioner, he is given some guidance as to the content of the regulations that he has a duty to promulgate and review. For example, the regulations must make provision for the training of those who conduct recombinant DNA research, for the establishment of institutional committees, and for personnel health monitoring systems. The bill is drafted in terms of mandatory certification rather than licensing, although the certification and enforcement procedures have almost exactly the same effect as the corresponding provisions of S.621. The only possible penalty is a suspension or revocation of the certificate or a maximum fine of $5,000 or both.

The bill was actually passed by the New York State Legislature on July 7, 1977, but the state's governor has refused to give his consent to it. In a

---

*See clause 4 and the Annual CEQ Report under NEPA. Although clause 4 refers to an annual report, the commission is to cease to exist 90 days after it makes its first annual report.
†Note that the inquiry is not limited to recombinat DNA techniques. See clause 8.

sense that was predictable. It can be argued that recombinant DNA research should not be regulated at the state level. The enactment of a uniform federal standard would seem to be preferable to uneven control based on differing state standards. Pressure from the scientific community can also have an important influence on the passage of such legislation, especially when federal preemption is precluded. In fact clause 9 specifies that the commissioner may promulgate regulations that are more stringent than those that are drawn up under the authority of federal law as long as there is no conflict with federal standards.

Two days after the emergence of the New York State bill the California Biological Research Safety Commission Bill was introduced into the California Assembly.[54] It adopted a strict liability approach in relation to injury to persons or property but, unlike S.621, it attaches strict liability only to injury that is caused by researchers engaged in "hazardous biological research" (clause 1797). This is defined as any "research, study, investigation or experiment which employs organisms or infectious agents which are capable, or can be rendered capable, of causing widespread serious harm, directly or indirectly, to the health of a substantial population of humans or to the natural environment" (clause 1782). It is possible to argue that the all-embracing strict liability proposed in S.621 was preferable in the sense that it is straightforward and certain in effect. The California approach does constitute a recognition of the fact that not all recombinant DNA research is capable of causing widespread, serious harm, but that distinction will be of little comfort to those who suffer loss. Unfortunately, the bill, which if enacted was to cease to have effect on January 1, 1981, was rejected by the California Assembly.

On March 8, 1977, a second bill was introduced into the U.S. Senate.[55] This bill represents an attempt to implement the NIH guidelines as regulations that would be administered by the secretary of HEW. The secretary could, however, prescribe more stringent regulations. In addition, the bill provides for the establishment of an interdisciplinary recombinant DNA study commission with a life of 27 months. This commission would make a comprehensive study of the ethical, social, and legal implications of recombinant DNA research. Nevertheless, it would have a scientific bias and six of its thirteen members would have to have been engaged in recombinant DNA research. The bill embodies the best features of S.621 and H.R.4232. It also gives the citizen, as well as the attorney general (at the request of the secretary), the right to seek an injunction against the continuation of recombinant DNA research. Such rights are often meaningless when they arise in the context of expensive litigation, but the bill has recognized that expense can often be a prohibitive factor and accordingly a plaintiff can be awarded all reasonable litigation costs, including the costs of experts, if the court determines that the plaintiff's action served an "important public purpose" (clause 10) or if it affords the plaintiff the relief sought in substantial measure. Thus even an unsuccessful litigant can receive his costs from the defendant.

The House responded to S.945 by introducing the Recombinant DNA Research Bill, which seeks to amend the Public Health Service Act.[56] Clause 473 is the most novel clause. It places a limitation on the number of facilities in which P4 research can be carried out. The secretary of HEW cannot designate more than ten laboratories as P4 facilities. Theoretically, this should lessen the likelihood of an accident. It may also represent a tentative first step toward the prohibition of all P3 and P4 research that is not conducted in a single national laboratory.

On April 1, 1977, Senator Kennedy introduced the Recombinant DNA Regulation Bill.[57] It provided, inter alia, for the registration of recombinant DNA projects and for its own expiration within five years of enactment. One of the most controversial provisions in the bill was clause 10, which permitted state and local governments to impose their own standards if those standards were at least as strict as comparable federal ones. After strong lobbying from the scientific community, Senator Kennedy introduced a substantially amended version of S.1217 on July 22, 1977.[58] State and local governments would be able to adopt more stringent regulations only if they were shown, on application to the secretary of HEW, to be "relevant and material to the health and environmental concerns" of the locality. This was still not viewed as a satisfactory formula and an amendment to S.1217 was subsequently introduced into the Senate.[59] Clause 484(2) provided that a local requirement that is more demanding than its federal equivalent cannot be enforced unless the secretary is satisfied that it is "necessary to protect health or the environment and is required by compelling local conditions."*

While S.1217 was being debated in the Senate, two recombinant DNA research bills had been placed before the House of Representatives. Both bills were introduced by Congressman Rogers within the period of a month. The first is significant for its $50,000 penalty provision, which clearly constitutes a more realistic attempt to control companies.[60] The bill is also notable because of the inclusion of a clause that relates to the disclosure of information acquired in the course of the enforcement of the act (see clause 480). If such information is within one of the exempted categories in the Freedom of Information Act,[61] it is not to be disclosed other than to officers whom the bill authorizes to receive it. However, there are certain exceptions as, for example, in situations in which disclosure is necessary for specific

---

*On September 27, 1977, Senator Kennedy announced that he no longer supported S.1217.

law enforcement purposes or to protect health or the environment against an unreasonable risk of injury (clause 480B[2]).

The new year began with the introduction into the House of yet another bill designed to control recombinant DNA research. The Recombinant DNA Safety Assurance Bill 1978 is concerned with the establishment of a commission to investigate the dangers involved in recombinant DNA research.[62] In that sense it does not represent a significant advance over the Commission on Genetic Research and Engineering Bill 1977, which had been considered by the House ten months previously. The second part of the bill simply relates to the implementation of the NIH guidelines in the form of statutory regulations.

Congress reconvened in September 1978 and if the bills which are before the House and the Senate are not brought up for debate before the end of the session they will expire with the 95th Congress. As one commentator has noted:

> Whether or not Congress passes a Bill before the end of the session depends not so much on the scientific and technical questions involved, or indeed on the social issues that arise immediately from them, but rather on how strong a political pawn it can be in the intricate power game that is played out on Capitol Hill.[63]

To a large extent, that power game is influenced by the views of a sizable group of scientists who have conducted a vigorous campaign against recombinant DNA legislation. Their attitude is reflected in the following statement:

> During 1977 the scientific community escaped a threat to the freedom of inquiry in the form of harsh legislation. The ostensible target was alleged hazards of recombinant DNA, but the objectives of some of the proponents were broader. The escape from restrictive legislation may prove to be only temporary. Last year Congressional action was delayed in part as a result of extremely effective lobbying by scientists . . . . If biologists relax the battle could be lost.[64]

Successive legislative compromises have not quelled the fears of these scientists. They are concerned that "measures now under consideration by Congressional, state and local authorities will set up additional regulatory machinery so unwieldy and unpredictable as to inhibit severely the further development of this field of research."[65]

It is clear that many scientists perceive the consequences of losing the "battle" as more real and immediate than does the general public. However, those consequences are short term and trivial when compared with the potential consequences of victory by those who cannot see beyond freedom of scientific inquiry. The widest possible public interest must prevail. It would be unfortunate if it were to be clouded by the supposedly subservient political process.

THE UNITED KINGDOM

In the United States, governmental interest in genetic research was stimulated in 1971 when James Watson delivered an address on genetic engineering to the Science and Astronautics Committee of the House of Representatives. That year was also an auspicious one for geneticists in the United Kingdom. It marked the establishment of a series of working parties on genetic research. The reports that they produced culminated in the promulgation of the world's first set of statutory regulations for the purpose of controlling genetic engineering experiments.[66]

In 1971 a study group sponsored by the British Association for the Advancement of Science published a report on the implications of advances in genetics and microbiology,[67] which was followed by several papers that represented attempts to keep pace with developments in these fields.[68] This interest in microbiological research was heightened in 1973 when there was an accidental release of smallpox virus from a laboratory in London. The virus was carried by one of the researchers at the laboratory and resulted in two deaths before the outbreak was contained.* A Committee of Inquiry immediately began to investigate the incident; all the evidence was heard in public and a full report was presented to Parliament in June 1974.[69] In addition, a working party, which was given the task of reviewing the laboratory use of dangerous pathogens, was established. The working party, which was chaired by Sir George Godber, recommended that a voluntary system of control should be coordinated by a Dangerous Pathogens Advisory Group (DPAG), which would be comprised solely of individuals who had themselves handled dangerous pathogens and who were familiar with current techniques and developments in that area.[70]

The public concern that had been aroused by the events of 1973 and the disturbing evidence contained in the report of the Committee of Inquiry coincided with the publication of the Berg letter in *Nature*† and led to the establishment, on July 26, 1974, of another working party with the following terms of reference: "To assess the potential benefits and potential hazards of experimental manipulation of the genetic composition of micro-

---

*An even more alarming incident occurred in Birmingham in August 1978. See the Introduction, note 1.

†The letter called for a moratorium on certain types of recombinant DNA research.

organisms . . . ." The working party, which was under the chairmanship of Lord Ashby, focused on recombinant DNA research and made only a passing reference to protoplast fusion techniques, which, it warned, should be carefully watched by the scientific community.[71] The working group recommended that the research be allowed to continue in view of its potential benefits to society. On the other hand it stressed the need for physical and biological containment measures and the training of laboratory personnel. In addition, the establishment of a central advisory service, which would offer guidance and information to researchers, was proposed. Throughout the report the distinction between the existence of risks to laboratory workers and risks to the general public was emphasized and that emphasis may have proved to be a significant factor in the recent decision to control genetic engineering by regulations under the Health and Safety at Work Act 1974.

The Ashby Report, and to a lesser extent the Godber Report, had been framed in uncomplicated but imprecise terms and it was evident that a further document would be necessary for the purpose of defining the details of control. Accordingly, another working party was established. This group, under the chairmanship of Sir Robert Williams, was asked:

> To draft a central code of practice and to make recommendations for the establishment of a central advisory service for laboratories using the techniques available for such genetic manipulation, and for the provision of necessary training facilities;
> To consider the practical aspects of applying in appropriate cases the controls advocated by the Working Party on the Laboratory Use of Dangerous Pathogens.[72]

The working party divided recombinant DNA experiments into four categories in terms of the risks involved and devised a code of practice for the conduct of experiments in each of the four groups. The emphasis differed from that in the U.S. National Institutes of Health guidelines in the sense that greater stress was placed on the need for trained personnel and there was also a strong bias toward physical rather than biological containment. Nor was it suggested that any type of experiment should be totally banned. On the other hand the proposed review structure was quite similar to the U.S. one. Researchers were asked to submit details of their experiments to a central Genetic Manipulation Advisory Group (GMAG), which can be equated with the NIH Advisory Committee. The GMAG would then advise the researcher on the category into which the work fell and on the relevant code of practice. The details of experiments were also to be submitted to the Health and Safety Commission (HSC), which would in turn receive technical advice from the GMAG. It was suggested that at the local level, every laboratory conducting genetic manipulation experiments should have an expert safety committee (equivalent to the local Biohazards Committee in the United States) and a Biological Safety Officer who would be answerable to the administrative head of the department or research establishment concerned.

The Williams Report also considered the possibility of establishing liaison between the GMAG and the Dangerous Pathogens Advisory Group, which had been set up after the publication of the Godber Report. It was felt that the DPAG should not be given the task of monitoring genetic manipulation experiments on the basis that

> the control of dangerous pathogens involves the application of well-known precautions against a small number of easily identifiable and well characterised agents. With genetic manipulation on the other hand the hazards and the precautions to be taken will depend on the detail of the experiment being undertaken.[73]

For the very reason that the dangerous pathogens are relatively few in number and easily identifiable, it can be suggested that existing DPAG functions should be combined with the oversight of genetic manipulation experiments.[74] This would avoid the need for time-consuming liaison between two groups with the same aims and similar procedures for controlling experiments that could involve agents classifiable both as dangerous pathogens and genetically engineered organisms. In practice, however, two different groups of experts with their own support staff would be likely to evolve even if DPAG and GMAG functions were discharged in the same building under the auspices of a single administration. On the other hand it is arguable that legislation in this area should cover conventional experiments with dangerous pathogens as well as genetic engineering experiments. Hence, a single empowering statute could spawn both a GMAG and a DPAG so that the fastidious control that would result from the utilization of two groups could be achieved with a minimum of overlap and a maximum of efficiency.

The Williams Report considered the possibility of passing a statute for the purpose of controlling genetic manipulation research. Control by means of existing legislation had already been discussed by the Working Party on the Laboratory Use of Dangerous Pathogens. That group was of the opinion that voluntary control measures should be backed by legislation and pointed out that the Factories Act 1961, the Diseases of Animals Act 1950, the Medicines Act 1968, and the Health and Safety at Work Act 1974 could conceivably be used for the purpose of controlling dangerous pathogens. The Williams Committee referred only to the Health and Safety at Work Act. It noted that limited control could be achieved through the use of existing provisions of that act in the sense that the Health and Safety Executive is given wide powers of inspection to enable it to ensure compliance with statutory requirements concerning the safety of workers and the general public.[75] On the other hand the protection of the act does not extend to plant and animal populations.

A specific statute directed toward compulsory consultation with the GMAG was tentatively suggested and the possibility of licensing laboratories received a passing mention.[76] The practical problems involved in drafting and administering a statute for the purpose of controlling a fast-developing science were also recognized and for that reason the committee proposed the promulgation of a set of statutory regulations that could be amended as new techniques emerge.

In December 1976 the recommendations contained in the Williams Report were acted upon. The first step was the establishment of the GMAG, which was attached to the Department of Education and Science. As had been suggested, its membership was not wholly scientific.[77] Eight scientists and medical experts* were joined by five representatives with a public interest, four employee representatives, and two managerial representatives.†
In accordance with the Williams proposals, consultation with the GMAG was, in the first instance, on a purely voluntary basis so that a decision on legislation could be deferred until after there was an opportunity to evaluate the operation of a voluntary system.

In its first year of operation the group gave advice on 102 proposals that were submitted from 27 centers.[78] The projects, which involved a total of 236 researchers, were categorized in terms of a fourfold classification system. These categories, ranging from I to IV, can be equated with the NIH containment levels, although category III experiments in the United Kingdom require more stringent precautions than their P3 equivalents in the United States.‡

The report is deficient in the sense that it does not deal with the question of whether or not the group's advice was followed by those who submitted proposals. The only suggestion that is made in that context is contained in a single sentence: "We propose to ask each proposer, a year after his proposal was received, to let us have a note setting out anything that might contribute to the accumulation of case-law and indicating the state of progress of the work in question."[79]

---

*These include three members of the DPAG, one Imperial Chemical Industries employee, and two consultants from two other corporations.

†Liaison with government departments and the Health and Safety Executive is maintained by nine "assessors" seconded from the relevant departments.

‡It is interesting to note that only four proposals received the category IV classification.

The need to minimize the delay that is involved in processing proposals is recognized in the report. In this context it is clear that the same factors that played an important part in the University of California and Harvard incidents also operate in the United Kingdom. The competitive nature of the work lends it an urgency that should not be slowed by inefficient administration. In fact, the vast majority of the proposals that were received by the GMAG were returned, with advice, in less than five weeks from the date of receipt.

On May 22, 1978, shortly after the publication of the GMAG's first report, the Williams recommendations were implemented in the form of the Health and Safety (Genetic Manipulation) Regulations 1978. The regulations, which were promulgated under section 15 of the Health and Safety at Work Act 1974, replaced a voluntary system of control with a mandatory one.* The three working party reports that had preceded the regulations were formulated in private. Invited witnesses were heard at closed committee meetings and the public was not given an opportunity to make submissions. The regulation-making process was not clothed in the same secrecy in the sense that the public could comment on draft regulations during a three-month period that was set aside for that purpose. Unfortunately, but predictably, the vast majority of comments that were made were submitted by scientists, many of whom were motivated by a desire to continue with projects that, in some cases, had already been in progress for several years. Even those researchers who were not involved in genetic engineering were justified in feeling threatened by legal intervention in genetic research in the sense that governmental intervention in the conduct of genetic engineering experiments would create a disturbing precedent that would have ramifications in other areas of scientific research. Accordingly, strong pressure from the scientific community[80] resulted in a dramatic narrowing of the definition of genetic manipulation that had been contained in the draft regulations.† The definition in regulation 2(1) is now confined to the

> formulation of new combinations of heritable material by the insertion of nucleic acid molecules, produced by whatever means outside the cell, into any virus, bacterial plasmid or other vector system so as to allow their incorporation into a host organism in which they do not naturally occur, but in which they are capable of continued propagation.

---

*The regulations came into effect on August 1, 1978.

†The Health and Safety Executive published the draft regulations as a consultative document that was circulated to all the relevant organizations. The broad definition in the consultative document related to "any activity intended to alter, or likely to alter, the genetic constitution of any genetic micro-organism."

The regulations are relatively straightforward. They require every person who is conducting or about to conduct genetic manipulation research to notify the GMAG and the Health and Safety Executive of their intention to carry out that research.* Regulation 6 gives the Health and Safety Executive, in conjunction with the GMAG, a discretion to exempt any person or class of person from the notification requirements, but there is no guidance as to the factors that should be considered in the exercise of the discretion. Regulation 3 simply extends the definition of work in the Health and Safety Work Act to "any activity involving genetic manipulation." Similarly, regulation 4 modifies section 3(2) of the act so that nonemployed persons are treated as if they were self-employed.† The task of administering and enforcing the regulations lies with the Health and Safety Executive. The act imposes general duties, on those conducting experiments, to ensure that the health and safety of workers and the public at large are not endangered. In an attempt to ensure that those duties are complied with, a large inspectorate was established under the 1974 act. Section 20 endows the executive's inspectors with wide policing powers and sections 21 and 22 empower the inspectors to serve improvement and prohibition notices.‡ In spite of these provisions it is doubtful whether the executive is the appropriate agency for controlling this research. If it is to be successful it will require new inspectors who must be allowed to concentrate solely on genetic manipulation.

In its first annual report the GMAG expressed the view that, if legislation were passed, it would be essential to separate the enforcement from the advisory role. Accordingly, it was seen to be important for the GMAG to operate only in the advisory sphere, thus leaving enforcement entirely in the hands of the executive. Conversely, it can be argued that if the group that makes recommendations also enforces them, the likelihood of the application of sanctions in the event of a breach will appear to be a more real possibility. The complete system of control that would presumably have been embodied in a specific statute, dealing wholly with the control of genetic manipulation, would have been preferable to the patchwork control that is offered by the regulations.** There are a number of defects both in the act and the regulations. For example, the act

---

*The notice is to be given on a form approved by the executive. See regulation 5.

†The Health and Safety at Work Act 1974, section 15(3) (a). This subparagraph empowers the secretary of state for employment to make regulations that repeal or modify any existing health and safety provision.

‡There is a right of appeal to an industrial tribunal in respect of improvement and prohibition notices. See section 24.

**The regulations might not represent the final solution. A seven-member subcommittee of the Select Committee on Science and Technology may recommend new legislation. It is currently in the process of evaluating developments in genetic engineering research and reviewing the regulations. A one-day "teach in" for the subcommittee was held at the University of Bristol on June 15, 1978. See *New Scientist*, August 10, 1978, p. 389.

applies to nongovernmental as well as governmental research but it does not extend to the protection of plant or animal life. The regulations refer to the notification of "any activity involving genetic manipulation" but they contain no reference to activities that involve the use of the products of genetic engineering research.*

OTHER FOREIGN JURISDICTIONS

In February 1977 the Canadian Medical Research Council (MRC) released *Guidelines for the Handling of Recombinant DNA Molecules and Animal Viruses and Cells.* The guidelines are nonmandatory although they are effectively binding on agencies that are funded by the government. The guidelines are more stringent than their U.S. and British counterparts. They extend not only to *in vitro* recombinant DNA research but also to more conventional *in vitro* research involving animal viruses and cells. Their scope is reflected in the identification of six categories of physical containment (A to F) instead of the usual four.†

The supervisory structure is three tiered. Responsibilities are attached to the principal investigator, the research institution, and the MRC. As would be expected, the most stringent duties rest with the principal investigator. The primary responsibility of the research institute is to establish a biohazard committee of the type that operates in the United States and the United Kingdom. This committee is responsible for ensuring that staff are properly trained in safety precautions and that the MRC guidelines are complied with.

The MRC's functions include the dissemination of information, the modification of guidelines, and the supervision of the local biohazard committees. In fact the MRC has its own nine-member biohazard committee, which is the only recombinant DNA biohazard committee in the world that has a majority of lay members — only four members of the group are scientists. As yet little information is available about the effectiveness of the MRC system, although there is no evidence to suggest that the guidelines are not being observed or that the preponderance of lay committee members has had anything other than a beneficial effect on the control process. However, the lack of information is, in itself, an indictment on the MRC committee. Its role should not be limited to the dissemination of

---

*Note, however, that section 6 of the Health and Safety at Work Act relates to the duties of designers, manufacturers, importers, or suppliers of articles used at work.

†The guidelines also define three levels of biological containment that are practically identical to the British and U.S. standards.

scientific and safety information to the local biohazard committees, but should also extend to a duty to inform the public, both in Canada and overseas, of the way in which its advice has been handled at the institutional level.

The Australians have been content to delay the introduction of formal controls on genetic engineering research until they have had the opportunity to evaluate the systems of control that are already operating in other countries. Nevertheless, in 1976 the Australian Academy of Science issued a short report that describes five categories of risk and associated containment measures.[81] The academy is an independent organization that receives government support but that neither controls research nor distributes funds. It has set up a standing committee on recombinant DNA molecules that, as its name suggests, does not consider wider categories of microbiological hazard. The eight-man committee, which has no lay members, has sent detailed questionnaires to research institutes and has asked them to submit their proposals for comments. As in other countries that operate voluntary systems, control can be exercised only over government-funded experiments by the body that allocates funds. For that reason the Health Department acts as the monitoring agency.

The most innovative of the Australian proposals relates to the establishment of an ad hoc committee for the purpose of reviewing the hazards associated with *in vivo* recombinant DNA experiments. The committee is concerned with the use of mixed infections, the production of new pathogens, and experimentation with certain types of viral pathogens. The dangers inherent in *in vivo* research have been neglected by all the other countries that have considered controlling novel genetic techniques and it is difficult to understand why concern has, in the past, been focused solely on *in vitro* experiments. Hopefully, the Australian precedent will not be ignored overseas.

A slightly different approach has been adopted in France. In June 1975 the Délégation Générale à la Recherche Scientifique et Technique established two national committees to study the implications of recombinant DNA research. The ten-member Ethical Review Group was charged with the task of considering the ethical, philosophical, and legal implications of the research, while the Control Commission was to review *in vitro* recombinant DNA research proposals and recommend safety procedures.[82] In practice the commission operates as a central administrative agency. None of its 12 members represents industry, trade unions or the public. Like the Australian National Academy it has distributed a questionnaire to relevant research institutes and will use the results to establish a national registry of recombinant DNA research. The commission has also issued guidelines that are very similar to the NIH recommendations.

Irrespective of the country that is involved, there is a certain uniformity about existing proposals for voluntary controls. Perhaps that was predictable.

What is more surprising is that the countries that are engaged in this type of research have, almost without exception,* gone to considerable lengths to promulgate workable guidelines† and to appoint supervisory committees. A vast majority rely on the containment measures and classifications described in the United States and the United Kingdom. Possible exceptions to that proposition include Austria and Japan, which tend to rely almost solely on the NIH guidelines, and Denmark and Sweden, which favor the recommendations that were made in the Williams Report.[83] Many jurisdictions, including Belgium, Italy, Switzerland, Holland, Denmark, and Japan intend to establish national registers of recombinant DNA research projects.

The greatest divergence in approach is noticeable in the membership of local and national supervisory committees and the emphasis that is placed on their respective roles. In Israel there is heavy emphasis on local safety committees that recommend appropriate precautions for particular experiments to a national committee composed partly of members who specialize in the humanities. In Sweden the national committee is of key importance. The 11-member committee established by the Royal Swedish Academy of Sciences is comprised of representatives of the lay public, industry, the scientific community, the Research Council, and the Ministry of Health. Its task is to liaise with the agencies that distribute grants and to advise researchers. On the other hand the Swiss Commission for Experimental Genetics has no lay representatives — it is composed entirely of experts in the fields of biology and medicine.

Some countries, including Denmark, operate a two-tiered system of control at the national level. In such cases one committee usually has a purely scientific membership while the other has a mixed membership that includes non-specialist members of the public, experts in ethics, and, more rarely, legal advisers. In the Netherlands three levels of national control have been proposed. The Committee on Genetic Engineering, which has an entirely scientific membership, operates at the top level of control. It has proposed the establishment of a special supervisory commission that would consist of members of the public and legal, ethical, and biomedical experts. At a lower level a "control committee" would be required to check that the requirements of the Supervisory Commission had been complied with. The

---

*Although recombinant DNA research is currently in progress at the International Institute of Biochemistry and Biophysics at the University of Iran, there is no evidence of an advisory agency or a set of guidelines. Portugal and Greece propose to set up national committees in the near future.

†For example, four sets of draft guidelines were formulated in the Federal Republic of Germany and the responsibility for the preparation of the guidelines was transferred from the German Research Council to the Ministry of Research and Technology.

main role of the Supervisory Commission would be to grant Certificates of Permission for the conduct of recombinant DNA research. This type of rigid structure with specific provision for follow-up and enforcement is simply a corollary of the Dutch view that legislation to control recombinant DNA research is required.[84]

Holland is not the only country, apart from the United States and New Zealand, that is currently considering the introduction of legislation to control this research. Ireland, Norway, and West Germany are also in the process of examining the possibility of legal control. Little is known about restraints on research in the communist countries, but it seems that Yugoslavia is controlling recombinant DNA research by means of existing legislation that is similar to the Health and Safety at Work Act 1974 (U.K.).[85] The German Democratic Republic appears to have adopted a set of guidelines, although there is little information on the actual mechanics of control in that country.

As in every other country involved in this research, scientists in the USSR have voiced conflicting views on whether there is a need for control. In the context of genetic engineering on humans, one Soviet scientist has stated that

> Soviet biologists . . . believe the ethical side of the problem has probably been exaggerated. At the moment, molecular biologists are happy if they can transfer just one gene . . . . The possibility of manipulating the huge number of separate genes that would be necessary if one was meddling with man's characteristics, is such a complex problem that it clearly lies in the distant future . . . when such genetic engineerng becomes possible, society will be mature enough to overcome the possible dangers.[86]

Frolov has presented a different perspective. In referring to genetic engineering research he has noted that "voices are being raised with increasing frequency concerning the need for democratic control over scientific research in fields abutting on vital interests of man and mankind."[87] Nevertheless it would seem that nonscientists do not sit on the Soviet Committee on Recombinant DNA Molecules, which is chaired by Professor Bayev. In fact its members are seconded from the Ministry of Health and the State Committee for Science and Technology. The committee is responsible to the Academy of Science and the Council of Ministers and has formulated a set of guidelines that draw heavily on those published by the NIH. In January 1978 the committee was in the process of amending the guidelines with a view to adopting more stringent precautions than had been recommended by the NIH. However, the USSR does not share many of the problems that have caused, and are likely to continue to cause, difficulties in other countries. Guidelines do not need to be given the force of law if all research is dependent on government funds. Similarly, the issue of corporate control does not arise. Those who seek to implement control on genetic engineering research in New Zealand are in a less fortunate position.

NEW ZEALAND

The first significant attempt to control genetic engineering research in New Zealand was made by the Plant Physiology Division of the Department of Scientific and Industrial Research. In June 1975 the division established a three-member committee for the purpose of investigating the hazards involved in the use of novel genetic techniques. That committee adopted a set of informal interim guidelines that served as a reference point for plant physiology staff. In the same month Peter Bergquist, who 15 months later was to conduct New Zealand's first recombinant DNA experiments, warned the Medical Research Council of New Zealand (MRC) of the possible dangers involved in this research. In particular, he stressed the necessity for careful consideration of the Asilomar recommendations* and the controversy out of which they arose. The MRC responded promptly and in July 1975 requested Bergquist and George Petersen to form an MRC Advisory Committee to evaluate genetic manipulation research and make formal recommendations.† Several meetings followed in the period between July 1975 and March 1977.[88] The New Zealand Biochemistry Society Symposium on Genetic Engineering, which was held at Palmerston North in August 1975, provided Bergquist and Petersen with an opportunity to discuss their views with the three-member Plant Physiology Committee. In March 1976 the DSIR conducted seminars in Auckland, Palmerston North, Wellington, and Christchurch on the topic of the future applications of molecular biology to agriculture and horticulture.‡ September marked the beginning of a course of lectures on genetic engineering that was offered by the Auckland University Centre for Continuing Education. Up to this point meetings had largely been the preserve of scientists; however, in late 1976 and early 1977 the issues involved were placed increasingly before the public.

The first call for a public inquiry was embodied in an article in *The Listener*.[89] Three weeks later the New Zealand Association of Scientists

---

*See note 6.

†Bergquist conducts his experiments in the Department of Cell Biology at Auckland University. Petersen is engaged in research with novel genetic techniques in the Department of Biochemistry at Otago University.

‡As a result of this exercise a report entitled "Creative Thinking at Cellular Level" was produced but not published.

requested an urgent public inquiry into the implications of genetic engineering research.[90] In their letter to the prime minister, Mr. Muldoon, the association stressed that "pending this enquiry there should be a moratorium on all further genetic engineering in New Zealand." In addition to drawing the attention of the public to the genetic engineering controversy, this action may have been at least partially responsible for the establishment of the DSIR Advisory Committee on Novel Genetic Techniques late in February.

In March 1977, nearly two years after the appointment of the MRC Committee, a set of guidelines that had been formulated by the committee was adopted by the MRC. These guidelines are of vital significance as far as the control of genetic engineering research in New Zealand is concerned.[91] In essence the committee recommended both physical and biological measures and a combination of British and U.S. guidelines. However, there is a highly disproportionate emphasis on the Williams recommendations, which, in the view of the MRC Committee, are more stringent than the NIH guidelines.

The committee also recommended that the guidelines should be flexible and that every research program involving recombinant DNA research (with certain limited exceptions) should be submitted to a Recombinant DNA Advisory Committee for approval.[92] The approval procedure is of particular interest. The MRC Committee suggested the establishment of a three-tiered system of control. At the lowest level the "principal investigator" would collaborate with a biological safety officer who was to be a member of an Employer Safety Committee that would be formed by the organization in which the research was conducted. The biological safety officer would be charged with the task of ensuring that the researchers in his organization were complying with the MRC guidelines and that all staff members had been informed of the nature of any relevant experiments that were being conducted by their colleagues.[93] The next tier in the structure was to be occupied by a Recombinant DNA Advisory Committee comprised of scientists from the MRC, the government, and the universities. Its function would be to examine the research proposals that would be submitted to it by individual researchers and to make detailed recommendations on procedures for the conduct of the experiments in question. In order to ensure that it was able to discharge that function, the committee and any person coopted by it was to have the power to inspect facilities and to instigate investigations into laboratory procedures.

At the top level of the hierarchy of control there was a proposed Genetic Manipulation Advisory Committee that was to be composed of scientists, trade unionists, representatives of industry, the Department of Health, and the public. It was to bear the responsibility for a continuing review of developments in the field of recombinant DNA research and would receive advice from the Recombinant DNA Advisory Committee on the risks involved in individual experiments. The proposal is represented in the following figure.[94]

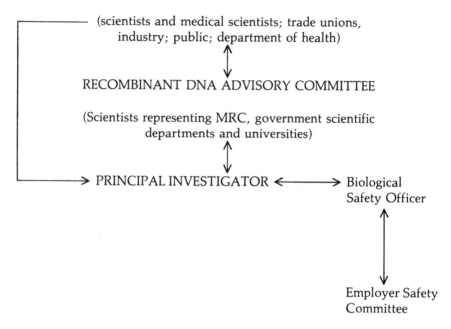

GENETIC MANIPULATION ADVISORY COMMITTEE

(scientists and medical scientists; trade unions, industry; public; department of health)

RECOMBINANT DNA ADVISORY COMMITTEE

(Scientists representing MRC, government scientific departments and universities)

PRINCIPAL INVESTIGATOR ⟷ Biological Safety Officer

Employer Safety Committee

In April 1977 the DSIR Advisory Committee, which had been established two months earlier, formulated a series of proposed modifications to the guidelines. The Advisory Committee accepted the proposed committee structure but suggested an extension of the scope of the guidelines in terms of the inclusion of somatic cell genetics and experiments with higher plants. These amendments were in fact incorporated into the guidelines and although the recommendations relate only to *in vitro* as opposed to *in vivo* recombinant DNA research,[95] the scope of the guidelines is now much wider than it is in other countries, with the possible exception of Canada, which regulates research involving experimentation with animal viruses and cells.

The first significant step toward the enforcement of these guidelines was taken on May 17, 1977, when the minister of science and technology, Mr. Gandar, directed several institutions to ensure that their laboratories were observing the modified MRC guidelines. Letters to that effect were sent to the Cawthron Institute, the Dairy Research Institute, the Meat

Research Institute, the Ministry of Forests, the Ministry of Agriculture and Fisheries, the Ministry of Health, Auckland University, Waikato University, Canterbury University, Otago University, Victoria University, Massey University, and Lincoln College. Although those organizations agreed to observe the guidelines, a certain amount of dissatisfaction was evident. For instance, in adopting its own "Regulations Governing Recombinant DNA Research,"* on September 19, 1977, the council of the University of Auckland noted that the regulations were dependent upon the National Genetic Manipulation Advisory Committee that had been recommended in the MRC guidelines. It expressed concern that neither that committee nor the proposed Recombinant DNA Advisory Committee had been established. Accordingly, the council resolved:

1. That the Minister of Science be informed that in accord with his letter of 17 May, the University has adopted regulations which depend for their implementation upon the establishment of a national Genetic Manipulation Advisory Committee, as envisaged in the recommendations of the Medical Research Council, and that he be urged as a matter of urgency to establish that Committee.
2. That as an interim measure, pending the establishment of the Genetic Manipulation Advisory Committee, that the University of Auckland regulations be implemented with the Medical Research Council Advisory Committee on Genetic Manipulation acting the role of the Recombinant DNA Advisory Committee, for the following category of experiment only:
Experiments between organisms that exchange genetic material in nature, and which comply with the requirements of Category I.

Apart from the fact that no other body had been established, the substitution of the MRC Advisory Committee for the Recombinant DNA Advisory Committee was presumably based on an assumption that category I experiments involve only minimal risks. More importantly, the DSIR was already recognizing the Bergquist and Petersen MRC Advisory Committee as the review body and that trend has continued up to the present. It is possible to argue that this recognition is justified in the sense that the Recombinant DNA Advisory Committee was to have a purely scientific membership drawn from the MRC, government, and the universities. Bergquist and Petersen satisfy both the first and the last of these three requirements. Bergquist appears to be the only researcher in the country who is currently involved in using recombinant DNA techniques in the truest sense and Petersen has a well-established reputation in the field of

---

*The Auckland University Regulations (1977) fill in some of the detail which is lacking in the MRC recommendations. They provide an interesting example of the way in which organizations could implement the MRC guidelines.

somatic cell genetics. On the other hand it would appear that wider scientific representation was envisaged. For example, the membership of specialists in fields like virology appears to have been contemplated by the MRC Committee.

It is clear that the minister should have attempted to establish the proposed supervisory structure, which offered the advantages of governmental, industrial, and public participation in the decision-making process, before he directed research institutes to observe the MRC guidelines. A realization of the inadequacies of the approach that was adopted may have led to the establishment of the government's working party on Novel Genetic Techniques,* under the chairmanship of Robin Irvine, the vice-chancellor of Otago University.† The working party was to report to the minister of Science and Technology and the minister of Health on the following issues:

1. Whether New Zealand should carry out research using novel genetic techniques at all
2. Whether action taken so far on interim guidelines on the use of novel genetic techniques is adequate and appropriate
3. Whether legislation is needed to regulate research using novel genetic techniques in the public and private sectors
4. Whether legislation, if required, should cover the wider question of microbiological hazards in research.

In August 1977 the minister of science and technology requested the DSIR Advisory Committee to prepare a paper on genetic engineering for the information of the members of the working party. The Advisory Committee prepared a useful background report that included a guide to the relevant literature and strong support for the modified MRC guidelines and the establishment of the proposed national committee for the continuing review of novel genetic techniques.[96] Some attention was also devoted to the possibility of enforcing the guidelines by means of regulations under an existing statute, as in the United Kingdom, or by the introduction of a new act. In this context the committee referred to the dangers of overregulation and unnecessary intrusions into freedom of inquiry.[97] It did not, however, attempt to define situations in which regulation would constitute an unnecessary intrusion.

---

*Novel genetic techniques include recombinant DNA, protoplast fusion, and somatic cell genetic research.

†Two other members were appointed, namely, H. N. Parton, a former vice-chancellor of Otago University and W. M. Hamilton, a former director-general of the DSIR. Bergquist and Petersen agreed to act as consultants.

The minister's request for compliance with the guidelines had been issued in May 1977. The machinery necessary for the implementation of the recommendations had still not been brought into existence by August 1977, when a certain amount of disquiet about the government's attitude to this issue became evident. Bill Wilson, a cell biology lecturer at Auckland University, suggested that a public inquiry into genetic engineering research would serve a more useful function than the working party.[98] In referring to Bergquist and Petersen and the MRC guidelines they had formulated, he stated: "I would suggest the two scientists have violated their social mandate by taking this decision, which is primarily a technical but also an ethical one." He went on to criticize the lack of legislation and machinery with which to enforce the MRC guidelines.

The Wilson article had its sequel in the House on August 17, 1977. Jonathan Hunt (M.P.) called for the government "in the interests of the people of New Zealand, to expand its Working Party into a public inquiry into genetic engineering in this country." In the press statement that followed, Hunt stressed that "genetic engineering is not a matter of scientific concern alone. It involves basic ethical and value judgments in which all New Zealanders have a right to be involved."[99] The Hon. Mr. Gandar simply replied that there was no point in having a public inquiry into genetic engineering as it was a highly technical field.[100]

On September 13, 1977, Richard Prebble (M.P.) gave notice of two questions for written answer. They concerned the precautions that the government was taking to control genetic engineering research.[101] In response, Mr. Gandar referred to the establishment of the working party but failed to comment on the implementation of the modified MRC guidelines.

The long-awaited working party report was finally released in August 1978. In some ways it represents a disappointing and retrograde step into the preguideline era. The first question that was addressed by the Irvine Committee was: "Should a regulatory authority prevent any research using novel genetic techniques being carried out in New Zealand?"[102] The Irvine Committee gave several reasons for its opposition to a complete ban. It noted that the class of novel genetic techniques is not limited to extremely or even moderately hazardous experiments, but also includes activities that potentially involve only slight risks. Accordingly, the benefits that this research could offer New Zealand were thought to be sufficiently important to mitigate against the imposition of a complete ban. Containment precautions could be utilized and the committee also expressed the belief that this type of research had already gathered too much momentum in New Zealand.

One of the most disturbing features of the report was the way in which the committee dealt with the issue of whether the action taken so far on the modified MRC guidelines was adequate and appropriate. The committee felt that the course of action that had been followed was appropriate but went on to make recommendations concerning measures that, it suggested, should be adopted in the near future. The first of these recommendations was that a five-member National Committee on Novel Genetic Techniques should be appointed to adjudicate on all proposed experiments that involve novel genetic techniques.[103] Perhaps "all" was the wrong word to use in view of the fact that five categories of novel genetic techniques on which the committee would not adjudicate are contained in a list that follows a description of three categories of experiment over which the committee would have jurisdiction.[104] However, it is interesting to note that experiments involving novel genetic techniques that were not specifically excluded would have to be referred to the committee before work could begin. The factors to be taken into account in the adjudication were the possible risks inherent in each experiment, the suitability of the laboratories in which the experiments would be carried out, and the training of the scientists involved. In addition, the committee would have a responsibility to continually update and review the guidelines, to advise local committees on the decisions that were made on individual projects, to publish all its determinations, and to submit an annual report of its work to the ministers of health and science and technology, to whom it would be jointly responsible.

The proposed National Committee is clearly the equivalent of the Recombinant DNA Advisory Committee, which had been recommended by the MRC, although the working party was more explicit about the composition of the committee than the MRC had been. It suggested that, although the National Committee should be composed entirely of scientists, the majority of the members should be scientists not directly involved in research with novel genetic techniques.[105]

Unfortunately, the similarity between the two reports ends at that point. The major departure in the working party report related to proposed committee structures. Only a single tier was envisaged at the national level. The committee did not attempt to justify its decision to ignore the Genetic Manipulation Advisory Committee that had been proposed in the MRC guidelines. Indeed the MRC recommendations on committees were not even mentioned.

The change is an important one. The Genetic Manipulation Advisory Committee, with its mixed scientific and nonscientific representation, would have provided an additional safeguard. It is possible that it might also have acted as an appellate body to which aggrieved scientists could have resorted in the event of dissatisfaction with the adjudication of the relevant local committee and the national advisory committee. Why then was the MRC proposal dropped? The answer to this question may be at least partly found in the submissions that were made to the working party by Bergquist and Petersen, who acted as consultants to the Irvine Committee and who, acting in their capacity as the MRC Advisory Committee, had formulated the MRC guidelines. They stated:

> In our earlier document, we suggested that the guidelines should be administered by a two-tier committee structure at the national level. This took the form of a committee of scientists advising a superior committee, consisting predominantly of laymen, which made the actual decision whether or not to allow a project to proceed. Upon reflection, we would wonder whether this is unnecessarily cumbersome. Since the Minister's ruling to DSIR and universities that the MRC guidelines should be followed in the meantime, bodies such as the DSIR and the University of Auckland have entrusted the oversight of these guidelines to their own Safety Committees. We would now like to suggest that a permanent arrangement along these lines might provide a workable alternative to our earlier proposal.
>
> A scientist on the staff, say, of the University of Auckland would, under this arrangement, submit his case to the small National Advisory Committee (of scientists) which would rule on the scientific aspects of the proposal, the risks, appropriate containment and so on, and pass its ruling on to the University of Auckland Council. The University Council would then make the decision whether to allow the work to proceed on its premises and would delegate responsibility for the oversight of the project and containment to its own Safety Committee, with the Biological Safety Officer performing the appropriate role specified in the guidelines. Enforcement would thus be nearer to home, the arrangement would still meet the need for the final decision to be in the hands of people other than scientists and there would be no need for a second national committee.[106]

The Irvine Committee accepted the submissions. It made only one minor modification to them in terms of a recommendation that local committees should report annually to the Health Department, through the National Committee on Genetic Techniques. Nevertheless the committee was obviously sensitive to the fact that its proposals would diminish the opportunities for public participation. Hence the lay representation on the local committees was emphasized, although it is noticeable that no weight had been attached to this factor when the MRC guidelines were formulated. At that time minimal lay representation at the local level was eclipsed by the power that would be held by lay members at the highest level of control.

The next question that was considered by the committee was whether legislation was needed to regulate research using novel genetic techniques in the public and private sectors.[107] The committee was of the view that there was not an urgent need to legislate to control this type of research in the public sector, largely because it was felt that compliance with national and local committee requirements could be enforced through funding constraints rather than by means of legislation. Private sector control was thought to be far more difficult. The possibility that multinational companies would conduct their research in New Zealand in order to avoid statutory control overseas was acknowledged, as was the risk of corporate noncompliance with nonmandatory controls. Problems with confidentiality and patents were also referred to but not discussed, and in fact the committee completely failed to grapple with the issues that are raised by the practical applications of the research and experimentation they sought to control. In the context of corporate control it was simply stated that the proposed two-tier committee structure "is obviously not appropriate, mainly for reasons of confidentiality." Instead of justifying that statement and attempting to propose an alternative structure, the committee preferred to recommend that "the Minister of Science and Technology should ask the DSIR to discuss with appropriate companies the philosophy behind this report, seek their adherence to the guidelines during the interim period before the introduction of legislation and ask them to keep the National Committee on novel genetic techniques informed of any developments."

The committee had very little to contribute on the mechanics of legislative control. Although passing references were made to the certification of vectors and the licensing of companies by the Health Department, no details were given.[108] Similarly, it was suggested that the National Committee on Novel Genetic Techniques should be serviced by the Health Department, but no attempt was made to elaborate on the type of assistance that would be involved. It was also concluded that the legislation should take the form of a new act rather than regulations or an amendment to an existing statute. Once again no reasons were adduced in support of this view.

In the next chapter an attempt will be made to provide the detail that was lacking in the report and also to answer the questions that were left unanswered by the committee.

NOTES

1. Panel on Science and Technology, Twelfth Meeting, "International Science Policy," Proceedings before the Committee on Science and Astronautics, U.S. House of Representatives, 93rd Cong., 1st sess., January 26, 27, and 28, 1971, pp. 336-66.
2. "Genetic Engineering, Evolution of a Technological Issue" (Washington, D.C.: U.S. Government Printing Office, 1972) and Supplemental Reports I and II.
3. Maxine Singer and Dieter Soll, "Guidelines for DNA Hybrid Molecules," Science 181 (1973): 1114.
4. Paul Berg et al., "Potential Biohazards of Recombinant DNA Molecules," Science 185 (1974): 303.
5. For a list of the participants, see Supplemental Report II, op. cit., Appendix 3.

6. Paul Berg et al., "Summary Statement of the Asilomar Conference on Recombinant Molecules," Proceedings of the National Academy of Sciences, June 1975, pp. 1981-84.

7. National Institutes of Health Guidelines for Research Involving Recombinant DNA Molecules, Federal Register 41, (July 7, 1976): 27902, 27913.

8. Ibid., p. 27913.

9. Ibid., pp. 27913-14.

10. Ibid., pp. 27917-20.

11. Ibid., p. 27911.

12. Ibid., pp. 27911 and 27920.

13. Gordon and Smith, "Have the Corporations Already Grabbed Control of New Life Forms?" Mother Jones Magazine, February/March 1977.

14. U.S. Senate, Labor and Public Welfare Committee, Subcommittee on Health, "Hearings on the Guidelines for Recombinant DNA Molecule Research," September 22, 1976.

15. The proceedings of the subcommittee are described in June Goodfield, Playing God: Genetic Engineering and the Manipulation of Life (New York: Random House, 1977), p. 154.

16. NIH Guidelines, op. cit., p. 27902.

17. 42 USC 4321.

18. Ibid., section 102(2) (c).

19. Draft Environmental Impact Statement on Guidelines for Research Involving Recombinant DNA Molecules, U.S. Department of Health Education and Welfare, National Institutes of Health, Federal Register, 41 (September 9, 1976): 38425; and National Institutes of Health Environmental Impact Statement, op. cit. On NIH Guidelines for Research Involving Recombinant DNA Molecules, October 1977.

20. The standing requirement has been interpreted very liberally and this approach has been affirmed in successive annual reports of the Council on Environmental Quality (CEQ), which plays an important role in administering the legislation. Barbaro and Cross cite the Third Annual CEQ Report, August 1972, in Ronald Barbaro and Frank Cross, Primer on Environmental Impact Statements (Westport, Conn.: Technomic, 1973).

21. 5 USC 553.

22. There is no indication in the act that Congress considered that NEPA would give rise to litigation or create enforceable rights. See "Environmental Impact Statements; New Legal Technique for Environmental Protection," Earth Law Journal 1 (1975): 15, 18.

23. Daily v. Volpe 326 F. Supp. 868 (1971).

24. Donald Fredrickson, director of the NIH, has also claimed that prompt publication of the guidelines served the public interest better than the publication of a prior impact statement that would have served only to delay the guidelines. See Colin Norman, "Laying the Guidelines Bare," Nature, 263 (September 9, 1976): 89.

25. Friends of the Earth v. Califano, 77 Civ. 2225.

26. 5 USC, App. I, section 5.

27. This statement is based on information that was kindly supplied by Richard Hartzman, attorney for the Friends of the Earth.

28. 447 Fed. Supp. 668.

29. See Mack v. Califano, op. cit., p. 669. Compare this attitude with the views that are expressed in Chapter 2.

30. Hanly v. Mitchell, 460 F. 2d, 649 (1972).

31. For further discussion of these problems, see Roger M. Leed, "The National Environmental Policy Act of 1969: Is the Fact of Compliance a Procedural or Substantive Question?" Santa Clara Law Review 303 (1975): 312.

32. National Institutes of Health Proposed Revised Guidelines for Research Involving Recombinant DNA Molecules, Federal Register September 27, 1977.

33. This comment was made at the public meeting by Peter Hutt, a Washington lawyer. See "Gene Splicing Rules: Another Round of Debate," Science 199 (January 6, 1978).

34. See Federal Register, July 28, 1978.

35. NIH Guidelines, July 7, 1976, op. cit.

36. See B. J. Culliton "Recombinant DNA; Cambridge City Council Votes Moratorium," *Science* 193 (July 23, 1976): 300.

37. Colin Norman, "Judgment of the People," *Nature* 265 (January 13, 1977).

38. See Culliton, op. cit., p. 300.

39. Nicholas Wade, "Harvard Gene Splicer Told to Halt," *Science* 199 (January 6, 1978): 31.

40. Described by Nicholas Wade in "Recombinant DNA: NIH Rules Broken in Insulin Gene Project," *Science* 197 (September 30, 1977): 1342.

41. Janet L. Hopson in *Smithsonian*, June 1977, as discussed in ibid.

42. *NIH Guidelines*, op. cit., p. 27905.

43. 42 USC 264.

44. On January 19, 1977, Congressman Ottinger introduced into the House a resolution that requested HEW to regulate recombinant DNA research under that provision. See H. Res. 131.

45. "Interim Report of the Federal Interagency Committee on Recombinant DNA Research," submitted to the secretary of HEW on March 15, 1977, and cited in Appendix J of the *NIH Environmental Impact Statement*, op. cit., Part II.

46. Public Law 91-596.

47. Public Law 93-633.

48. Public Law 94-469.

49. Public Law 88-206; Public Law 91-224; and Public Law 94-580.

50. Delbert S. Barth, "Environmental Protection Agency Submission to the Subcommittee on Recombinant DNA Research," 1977 (unpublished).

51. S.621, 95th Cong., 1st sess. The bill was introduced by Senator Bumpers.

52. H.R. 4232, 95th Cong., 1st sess.

53. Article 32-A, Senate Bill No. 4009 — Cal. no. 448, D: A Bill to Amend the Public Health Law in Relation to the Certification of Recombinant DNA Experiments, 1977.

54. California Legislature, No. 757. The bill attempted to add a tenth chapter to Division 2 of the Health and Safety Code.

55. The Recombinant DNA Standards Bill 1977, S.945, 95th Cong., 1st sess., Introduced by Senator Metzenbaum.

56. H.R. 4759, 95th Cong., 1st sess., introduced by Congressman Rogers.

57. S.1217 95th Cong., 1st sess.

58. Nicholas Wade, "Confusion Breaks Out Over Gene Splice Law," *Science* 198 (October 14, 1977): 176.

59. Recombinant DNA Bill, S.1217 Amendment No. 754, Senator Nelson, *Congressional Record* — Senate, August 2, 1977, p. 13312. It is also interesting to note that clause 4(1) of the bill specifically provides that sec.102(2) (c) of the National Environmental Policy Act, which relates to the preparation of environmental impact statements, will not apply to the promulgation of regulations under the act.

60. Recombinant DNA Bill 1977, H.R. 7418, 95th Cong., 1st sess., May 24, 1977. The second of these bills (H.R. 7897, introduced June 20, 1977) is not sufficiently different from its immediate predecessor to warrant discussion here.

61. Title 5, USC; section 552(a).

62. H.R. 10453, 95th Cong., 2d sess., January 19, 1978 (introduced by Congressman Staggers). If enacted, H.R. 10453 would cease to have effect after two years.

63. John Richards, in "Recombinant DNA in Focus," *Nature* (October 2, 1978): 14.

64. Philip H. Abelson, "Recombinant DNA Legislation," editorial, *Science* 199 (January 13, 1978): 135.

65. From the text of an open letter signed by 137 (75 percent) of the geneticists who attended the Gordon Conference on Nucleic Acids in 1977. It is ironic that the scientists who attended the 1973 Gordon Conference issued the first of the many warnings about recombinant DNA research that were to emanate from the scientific community.

66. The Health and Safety (Genetic Manipulation) Regulations 1978 (U.K), No. 752.

67. "Social Concern and Biological Advances," Report of a Study Group, British Association for the Advancement of Science, Publication 7412, September 1974.

68. See Alun Jones and Walter Bodmer, *Our Future Inheritance: Choice or Chance?* (London: Oxford University Press, 1974).

69. "Report of the Committee of Inquiry into the Smallpox Outbreak in London in March and April 1973," HMSO Cmnd. 5626, June 1974.

70. "Report of the Working Party on the Laboratory Use of Dangerous Pathogens" (Godber Report), HMSO Cmnd. 6054, May 1975. It is interesting to note that the Department of Health has appointed the chairman of DPAG, Sir Reginald Shooter, to lead an investigation into the Birmingham incident. As one commentator has already suggested, "it is certain that conditions in laboratories handling hazardous pathogens are less than perfect. The suspicion that the Department of Health is to some extent responsible for these lamentable inadequacies is difficult to dispel when its attempts to water down recommendations from its own advisors . . . are all too obvious. The Department does not have a clean record and cannot expect the public to accept without question the verdict of one of its chief advisors on his own work." See Lawrence McGinty, "Passive Immunity," *New Scientist*, September 7, 1978, p. 666.

71. "Report of the Working Party of the Experimental Manipulation of the Genetic Composition of Micro-organisms" (Ashby Report), HMSO Cmnd 5880, January 1975, p. 5.

72. From the "Report of the Working Party on the Practice of Genetic Manipulation" (Williams Report), HMSO, Cmnd. 6600, August 1976.

73. Ibid., paragraph 5.5.

74. Seventy - two dangerous pathogens were identified in the Godber Report, op. cit., Appendix I.

75. See Health and Safety at Work Act 1974, sections 20, 21, and 22.

76. Williams Report, op. cit., paragraph 5.13.

77. Ibid., paragraph 5.3.

78. "First Report of the Genetic Manipulation Advisory Group," HMSO Cmnd. 7215, May 1978. See paragraph 3.2, Table 1. The group was set up for a two-year period.

79. Ibid., paragraph 3.15.

80. For example, see Michael Ashburner, "An Open Letter to the Health and Safety Executive," *Nature* 264 (November 4, 1976): 2; and "Genetic Proposals; Reactions," *Nature* 264 (November 18, 1976): 209.

81. Gordon L. Ada, "Guidelines for Both Physical and Biological Containment Procedures for Work Involving Recombinant Nucleic Acid Molecules," unpublished.

82. For a more detailed description of the two groups see Assemblée Nationale, 3rd Séance du 17 Novembre 1976. During 1976 and 1977 the Commission reviewed and classified 50 proposals. See *NIH Recombinant DNA Technical Bulletin* 1, no. 2 (Winter 1978).

83. See the "First Report of the Genetic Manipulation Advisory Group," op. cit., p. 75.

84. See *Report of the Committee in Charge of the Control on [sic] Genetic Manipulation*, Royal Netherlands Academy of Arts and Sciences, March 1977, p. 28.

85. "First Report of the Genetic Manipulation Advisory Group," op. cit., Appendix IV, p. 77.

86. Vladimir Engelhardt as reported in "A Unique Plan for Soviet Molecular Biology," *New Scientist*, January 8, 1976, p. 53.

87. I. T. Frolov, "Research on Man, Genetic Engineering," *Voprosy Filosofii*, No. 7, (1975) U.S. JPRS 66307, December 5, 1975, pp. 83-95.

88. See "Novel Genetic Techniques in New Zealand," Report of the DSIR Advisory Committee, October 1977, pp. 12-14.

89. Bill Wilson et al., "Genetic Engineering — New Hope or New Horror?" The New Zealand *Listener*, February 12, 1977, p. 14.

90. "Inquiry Urged Into Genetic Risks," New Zealand *Herald*, March 7, 1977.

91. See the "Recommendations of the MRC Advisory Committee on Genetic Manipulation," 1977 (unpublished).

92. Ibid.

93. Considerable importance was also attached to training programs for laboratory staff. See ibid., p. 4.

94. See ibid., p. 5. These suggestions are similar to those that have been made in Denmark both in regard to the number of tiers involved and the membership of the committees. The committee also expressed the view that, despite the fact that there were no category III or IV containment laboratories in New Zealand, it would be more economical to conduct high-risk experiments in suitable laboratories in Australia, the United Kingdom, or the United States than to build an appropriate facility in New Zealand, although the adoption of the former alternative would give rise to difficulties with quarantine and transportation.

95. The committee expressly excluded experiments that fall within a wider category of microbiological hazard, that is, experiments in microbial and bacteriophage genetics that do not involve *in vitro* recombinant DNA techniques and that are conducted on organisms that do not naturally exchange DNA. See ibid., p. 3. The committee also suggested that certain classes of experiment should be prohibited altogether, see ibid., Appendix I.

96. "Novel Genetic Techniques in New Zealand," op. cit.

97. Ibid., p. 25.

98. "Genetic Engineering Work Would be Better Replaced by An Inquiry," *Evening Post*, August 11, 1977. Wilson has the full support of Bob Mann and the Environmental Defense Society in this matter.

99. *Evening Post*, August 17, 1977.

100. New Zealand *Herald*, August 19, 1977. The minister of health had expressed a similar view. See "Genetic Engineering — A Continuing Saga," 35 New Zealand *Science Review*, 35 (1978): 3.

101. House of Representatives Supplementary Order Paper, September 13, 1977. Compare a series of 11 questions that were asked by Leo Abse (M.P.) and answered by John Grant, the Parliamentary Undersecretary of state for employment. See Parliamentary Debates (U.K.) 642 (House of Commons) (June 15, 1978): 1111.

102. "Report of the Working Party on Novel Genetic Techniques," 1978, Parliamentary Paper, G21A. The committee rephrased this question, which had originally read: "Should New Zealand carry out research on novel genetic techniques at all?" The DSIR Advisory Committee had simply begun its discussion by stating that "It is . . . *presumed* [emphasis added] that New Zealand will wish to continue to have some competence in this area of work, and not to place a veto on all such experiments." See "Novel Genetic Techniques in New Zealand," op. cit., p. 23.

103. It was suggested that decisions on applications should be forthcoming within one month of the date on which they are received by the committee. See the "Report of the Working Party on Novel Genetic Techniques," op. cit., p. 7.

104. See ibid., paragraphs 6(a) to (h).

105. It was also proposed that the committee be given a small budget for site visits and training programs. See ibid., p. 5.

106. Peter L. Bergquist and George W. Petersen, "Submission to the Irvine Committee," 1978, unpublished.

107. At the New Zealand Law Society's Triennial Conference, which was held in Auckland five months prior to the publication of the working party report, McMullin, J. had stated that the problems of genetic engineering research were urgent and that "intervention by international action and national statute will be necessary." See "Conference Courier," August 28, 1978, p. 1.

108. "Report of the Working Party on Novel Genetic Techniques," op. cit., pp. 6 and 8. Another recommendation involved the publication of safety precautions in all journal articles concerning novel genetic experiments. See p. 8.

4

THE MECHANICS OF NATIONAL LEGAL CONTROL: SOME PROPOSALS FOR NEW LEGISLATION AND UTILIZATION OF EXISTING LAW

GENERAL PRINCIPLES CONCERNING THE ROLE OF THE LAW

The fact that there is a need to control the technology of genetic engineering has been established in the preceding chapters. The next issue that arises is whether legal control is more desirable than formal and informal nonlegal control. The proponents of an informal nonlegal approach point to the constraint of peer group pressure and, at a more formal level, to the sanction of a withdrawal of funds. However, peer pressure has not always been effective in the past.[1] It must compete with the attraction of the financial gain and acclaim that would be likely to accompany a successful but high-risk experiment.

An actual withdrawal of funds by a sponsoring agency should operate as an effective control measure, although that sanction is limited to government-funded laboratories. The same consideration has governed the effectiveness of nonlegal formal controls such as the NIH guidelines. In the absence of a condition in the original research grant, they do not bind and can be enforced only by a withdrawal of funds in the case of government-sponsored institutes.

There is a pressing need for the introduction of legal sanctions. In addition, legislation has a symbolic function that extends beyond the apparent force of its penalty provisions. The enactment of a statute is, in itself, an indication of the importance that society attaches to the area that is being regulated. Spigelman describes this as a "social denunciation" factor, but he warns that "all or nothing" remedies such as license revocation are not adequate, in the sense that legislators must acknowledge the wide variety of considerations that can affect the weight of the denunciation.[2] He points out that these considerations are also related to the importance of flexibility in the sanction and remedy structure. If a statute were passed specifically for the purpose of controlling genetic engineering research, the necessary degree of flexibility could be ensured by providing for variable and perhaps novel remedies and sanctions that could be tailored to fit the culpability and seriousness of the acts or omissions involved.

The law can be criticized as a mechanism of control on grounds other than its supposed inflexibility. For instance, it can be argued that legal intervention is unlikely to result in effective control when research has already gained considerable momentum and that lawyers and legislators should not attempt to control technologies that they do not fully understand.[3] It has also been claimed that the implementation of legal controls over one area of scientific research could lead to the control of other fields of biomedical research where regulation is said to be unnecessary in view of the fact that the risks are thought to be well known.[4] Similarly, there have been complaints about the cost that would be involved in enforcing and adhering to legal controls.

These arguments fail to convince. Although it is clear that nonscientists may not have the technical expertise to fully understand the new technology of genetic engineering, scientists labor under a similar handicap in relation to the moral, ethical, social, and legal implications of their work. This is not to suggest that neither of the two groups can become acquainted with the special skills of the other but rather that a partnership is necessary. A combination of aptitudes is required. For instance, in the context of new legislation for the control of genetic engineering research, scientists have an extremely important consultative role to play, although, in practical terms, consultants have already come disturbingly close to usurping legislative functions.[5]

Arguments that relate to the momentum and unstoppability of the research would be more persuasive if a complete ban rather than an attempt to regulate genetic engineering experimentation were contemplated (see Chapter 2). In the past regulatory rather than purely prohibitive controls have been successfully imposed on important, fast growing, and highly technical sciences.[6] It is also recognized that the factors that relate to a need for control will vary from case to case. Thus, legislative intervention in one area has not been followed by sudden legal interest in the control of other scientific techniques, although increased vigilance should not be unwelcome in any event. Similarly, the cost of compliance factor is insufficient as a justification for the absence of legal controls. The benefits offered by genetic engineering research are considerable as is the risk of extremely costly accidental damage. On the other hand the cost of compliance with safety precautions that would minimize the risks and maximize the benefits is relatively small.

The situation is somewhat different in relation to the common law. In that sphere it is more difficult for scientists to argue that interested members of the public should not be able to pursue existing remedies against researchers who have caused, or who are likely to cause, damage. The rights and remedies of the plaintiff form part of an accepted legal system and all that remains is for the litigant to fit the contours of the new technology into the old framework. That framework and any existing statutory controls must now be evaluated. If, as in the United States (see Chapter 3, first section), the existing structures do not provide an adequate network of control, a bill designed specifically for the purpose of controlling genetic engineering research should be introduced into the House of Representatives without delay.

EXISTING LAW

Prevention

Most of the existing means of preventive control have a statutory rather than a common law base. However if a nuisance action is brought, injunctive relief is available, although the court's discretion to grant an injunction will usually be exercised only if the interference complained of is serious and continuing. On the other hand it is possible for a plaintiff to seek a *quia timet* injunction. If granted, such an injunction would bring an end to an activity that, in the opinion of the court, is about to cause imminent and irreparable damage. Clearly, injunctions issued in the context of nuisance actions are not likely to provide an effective means of halting certain types of high-risk genetic engineering experiments. The courts would quite rightly be reluctant to suggest that the risk of damage resulting from such experiments is so high that the research should be stopped. Such a decision is obviously one that should be taken by the legislature. If the risks are sufficiently large, the government should step in and ban the experiments. In addition, the nuisance action has not developed in a way that lends itself to application to genetic engineering accidents that are likely to consist of isolated bacterial or viral escapes from a laboratory.[7] It could be argued that escaped microorganisms will replicate outside the laboratory and continue to cause damage in the community, but in such cases an injunction could relate only to the cessation of the original experiment. It would also be highly unlikely that the institute in question would continue an experiment under the circumstances, and there is no practical remedy in respect of the continuing replication and damage in any case.

The limited nature of existing injunctive remedies reinforces the view that the common law is a particularly deficient weapon for preventing damage. However, the Plants Act 1970 (N.Z.) is one of several statutes that could conceivably be used to prevent genetic engineering accidents. Section 2 of the act defines disease as "any unhealthy condition in any plant material . . . which may be caused directly or indirectly by any form of fungus, bacterium, virus or micro-organism; which may cause such a condition." The words "any form" would seem to cover pathogenic genetically engineered organisms. The act establishes a system of control that includes provision for quarantine,[8] the proclamation of plant disease emergencies,[9] the eradication of disease,[10] and restrictions on importation,[11] although it is interesting to note that the director-general of Agriculture "may for the purpose of scientific research or experiment permit the importation into New Zealand of anything not otherwise eligible under this Act for importation."[12] The word "anything" would, at the very least, appear to extend to the importation into New Zealand of genetically engineered organisms for the purpose of research, even if the imported organisms are capable of producing disease in animals as well as plants. The section is reinforced by section 31 (3), which provides that regulations made under the act may prescribe that it will be necessary to obtain a permit for the introduction into New Zealand of any bacterium, virus, microorganism, or any other thing.

The act, which binds the crown, is enforced by specially appointed inspectors who have limited rights of entry onto land and premises.[13] In addition offenses against the act are set out in section 28 and a maximum penalty of $500 is prescribed.

The Forests Act 1949 (N.Z.) also makes provision for regulations governing the control and eradication of diseases that could affect trees, forests, or forest products.[14] The definition of disease, which is slightly wider than that in the Plants Act, appears to cover pathogenic genetically engineered organisms. It relates to "any form of fungus, bacterium or virus, or any living stage of any invertebrate animal, which may directly or indirectly injure or cause an unhealthy condition in any tree or any other plant. . . ."[15]

Although the definition of disease in the Plants Act and the Forests Act can be said to extend to certain types of genetically engineered organisms and viruses, the acts and indeed the definitions were drafted before genetic engineering techniques were developed and thus it is hardly surprising that the acts can be applied only to the control of genetic engineering experiments in certain limited circumstances. For example, the Plants Act offers an opportunity for control in the context of the importation of microorganisms into New Zealand,* but beyond that it is simply not adequate to cope with a situation for which it was not designed. The Forests Act suffers from the same defects and, like the Plants Act, it does not envisage the protection of animal as opposed to plant life. The same problems would exist in relation to the utilization of the Noxious Plants Act 1978 (which repealed the Noxious Weeds Act 1950). Under sections 18 and 19 of the act certain types of genetically engineered organisms could be declared to be noxious plants, but it is clear that those who drafted the act did not consider the control of laboratory pathogens and hence its enforcement provisions offer little hope of assistance.

As might be expected, the disease control measures in the Animals Act 1967 follow a similar pattern to those in the Plants Act. However, a different approach to the definition of disease has been adopted. A disease is declared to be any disease specified in the first or second schedule of the act and an organism is defined as

> any protozoan, fungus, bacterium, virus, or any other organism or micro-organism, being one which if living is capable of causing or transmitting any disease as defined in this section or any other disease affecting animals or, if dead, was so capable when living; and includes any culture, subculture, or other preparation of any such protozoan, fungus, bacterium, virus, organism or micro-organism.[16]

Once again the act does not contemplate the control of factory workers and researchers, although its definition of "organisms" would certainly include some of the products of their work. The Health Act 1956 is of greater interest in the sphere of enforcement. It provides for the establishment of a Board of Health, Advisory and Technical Committees, Medical Officers of Health, and Local Body Health Inspectors.[17] However, the act is basically concerned with the control of sanitary conditions in the community.† Infectious diseases are specified in parts I and II of the first schedule to the act and that specificity could be extended, by amendment, to different categories of pathogenic genetically engineered viruses, fungi, bacteria, and organisms. Such an amendment would, however, be unwise. The Health Act, like all the other statutes that have been reviewed in this chapter, does not provide a control mechanism that is directed toward the special problems that are involved in controlling genetic engineering research.

---

*For example, see section 18, which empowers inspectors to direct the reshipment, treatment, or destruction of any diseased plant material that is imported into the country.

†For example, see section 29, which defines nuisances for the purposes of the act.

The Clean Air Act 1972 is a more appropriate statute on which to base control.[18] The definition of air pollution in section 2 could include genetically engineered microorganisms. An air pollutant is defined as

> anything of harmful, odorous, or offensive character, in such a form that it can be carried in the atmosphere and in particular, but without prejudice to the generality of the preceding words, includes smoke and any gases, fumes, mists, or dusts containing any substance specified in the First Schedule to [the] Act.

Although the latter words are expressed to be without prejudice to the opening words of the definition, the classes of air pollutants that are specified in the first schedule demonstrate an exclusively inorganic bias that is reflected in the provisions of the act.[19] On the other hand the act has many desirable features that could be used in the control of genetic engineering research if a category of genetically engineered microorganisms were included in the first schedule. It is concerned not only with the control of air pollution but also with its prevention and for this purpose a Clean Air Council was established.[20] Section 23 relates to the licensing of scheduled processes and the relevant processes are specified in parts A, B, and C of the second schedule to the act. The superficial attraction of describing genetic engineering techniques as scheduled processes begins to wane after an examination of existing scheduled processes. Most relate to combustion although the schedule includes

> any industrial chemical processes, including electro-chemical processes having as a product or by-product or emission any substance that can cause air pollution, including any processes used in . . . synthesis or extraction of organic chemicals, including formulation of insecticides, weedicides, plant hormones and like toxic or offensive organic compounds.

Under section 27 the licenses that are granted in relation to the premises on which scheduled processes are carried out must be registered and the duty to enforce the act's registration provisions is placed on local authorities.[21] Penalties for noncompliance with the act range up to $5,000, an upper limit that may be inadequate in the context of corporate control but that is certainly more appropriate than the small penalties in the Plants and Animals Acts.[22]

There is little doubt that an attempt to control genetic engineering research under the provisions of the Clean Air Act could be made. However that course of action should be avoided. The act does not offer a unified approach to genetic engineering techniques.[23] Like the other statutes that could, on a strained interpretation, be said to extend to genetic engineering, it recognizes neither the nature nor the significance of the techniques that are involved.

Compensation

In the absence of administrative action, most existing common law remedies, with the notable exception of the equitable injunction, redress the balance as between plaintiff and defendant *after* damage or a breach of duty has been caused. In that sense the common law offers opportunities for compensation rather than control. Compensatory remedies are based on proof of causation. The plaintiff must prove, on the balance of probabilities, that the defendant's acts or omissions were the cause of the damage or interference that forms the subject of the action. Proof of fault or damage will not always be necessary, but the plaintiff cannot escape the burden of proving causation and it is this onus that may form a practical bar to recovery of damage in the case of claims relating to genetic engineering.* There are also likely to be problems with the question whether the plaintiff was foreseeable. These issues can be best illustrated in the context of a discussion of each of the relevant common law actions.

The action in negligence could be used by a plaintiff who had sustained property loss as a result of a genetic engineering experiment or process. Similarly, if personal injury were sustained and such injury could not be said to constitute personal injury by accident within the meaning of section 2 of the Accident Compensation Act (N.Z.) then a common law action could be brought against the negligent party. The first problem that could arise in the context of a cause of action based on negligence would be the difficulty of establishing that either genetic engineering researchers or those who use the products of genetic engineering research owe a duty of care to the public or, in other words, that they are under a legal duty to conform to a standard of conduct for the protection of others against unreasonable risks. In practical terms it is difficult to distinguish this issue from that of the foreseeability of the damage that is sustained by a particular plaintiff. In Lord Atkin's terms,

> you must take reasonable care to avoid acts or omissions which you can reasonably foresee would be likely to injure your neighbour. Who, then in law, is my neighbour? The answer seems to be — persons who are so closely and directly affected by my act that I ought reasonably to have them in contemplation as being so affected when I am directing my mind to the acts or omissions which are called in question.[24]

On its face this vague test could lead to the conclusion that genetic engineers owe a duty of care to members of the public who sustain loss as a

---

*For the purposes of this discussion it is presumed that the acts involved are accidental as opposed to intentional.

result of their experiments. However, Lord Atkin's dicta are narrowed by policy considerations. As Fleming points out, "the duty issue involves an exercise of the specific creative function of the judiciary in controlling the area of legal responsibility for negligence, having regard both to the nature of the interests infringed and the type of conduct complained of."[25]

The policy factors that influence the so-called creative function assume alarming proportions when the conduct complained of is personal injury or property damage caused by genetically engineered microorganisms. For example, a single accident with *E. coli* bacteria could potentially result in widespread and serious personal injury. Similarly, a genetically engineered fungus such as the one manufactured at the DSIR Plant Physiology Laboratory at Palmerston North could, if it escaped from the laboratory, cause severe damage to the forestry industry. The enormity of the damage that could be involved might mitigate against the establishment of a duty in the genetic engineering context. It can also be argued that in the event of widespread damage the individuals who suffer loss would be the best loss bearers. An individual researcher or his institution might not be able to compensate more than a very small proportion of those who are injured by his conduct, although the possibility of insurance may temper the policy considerations that are involved in this area.*

In the past, judicial creativity in terms of controlling the scope of legal responsibility for negligence has not been exercised sparingly. In *Donoghue v. Stevenson*[26] the court was reluctant to extend liability for negligent acts to the manufacturer of a defective product. However, a willingness to identify a duty to large classes of plaintiffs is not now uncommon. For instance, liability for negligent words was initially restricted on the basis "that words are more volatile than deeds, they travel fast and far afield, they are used without being expended."[27] Nevertheless, even in that area responsibility has been acknowledged by the court. Perhaps microorganisms can be equated with fast-traveling, volatile words. However, the comparison must end at the metaphor and the potentially large class of plaintiffs in each case.

In *Weller v. Foot and Mouth Disease Research Institute*[28] a claim by cattle auctioneers against a research institute that accidentally released a foot and mouth virus failed because the auctioneer's loss was incurred as a result of the closure of the cattle market and not because of the destruction of stock that had never belonged to the auctioneers. † Hence the decision cannot be cited as authority for the proposition that the categories of negligence do not extend to the damage caused by escaped laboratory

---

*For a discussion of the arguments surrounding the establishment of a special compensatory fund, see the last section of Chapter 5.

†The plaintiffs argued the case mainly on the basis of the negligence of the defendants.

viruses. In fact the case contained several interesting statements that, although they are *obiter,* could influence the courts toward allowing plaintiffs to recover losses that are caused by the negligent release of genetically engineered organisms.

For example, Widgery, J. acknowledged that the defendants owed a duty of care to cattle owners in the district. He stated: "In the present case, the defendants' duty to take care to avoid the escape of the virus was due to the foreseeable fact that the virus might infect cattle in the neighborhood and cause them to die."[29] However, he also issued the following warning: "The categories of negligence never close, but when the Court is asked to recognise a new category, it must proceed with some caution."[30] That statement is undoubtedly representative of prevailing judicial attitudes concerning claims relating to damage that is caused by processes in respect of which actions have never been brought in the past, although it must be noted that Widgery, J. appears to have been commenting on the category of consequential as opposed to direct loss and not on the possibility of recovery for damage caused by escaped viruses.

Even if the courts are reluctant to explicitly deny that those who use genetic engineering techniques owe a duty of care to the public, they might be prepared to limit liability in more subtle ways. If the duty of care is acknowledged, the question whether or not there has been a breach of that duty, in the sense that the researcher has fallen below the standard required by the law, does not provide the court with a useful opportunity to limit liability. It is already well-established that failure to comply with non-mandatory professional safety codes can be used as highly persuasive evidence of negligence.[31] Noncompliance with the MRC guidelines would seem, at the very least, to point toward a breach of the researcher's duty of care.

The doctrine of causation potentially constitutes a far more powerful practical limitation on recovery in the context of the technology of genetic engineering. In a negligence action the plaintiff must prove, on the balance of probabilities, not only that he has sustained injury to his person or property but also that the defendant's negligent actions caused that damage. Several different types of negligent conduct can be envisaged. For example, there would not be a great deal of difficulty in proving that a sudden, dramatic reduction in the quantity of oil that was stored in a refinery was caused by the negligent release of recombinant *pseudomonas* bacteria.* The causal link could be established by evidence of *pseudomonas* in depleted storage tanks. There would, however, be more extreme difficulties in

---

*The DNA of these organisms has been recombined in order to produce an oil-consuming capacity.

proving a causal connection in the case of an outbreak of illness that was alleged to have been caused by the escape of a genetically engineered virus.[32] The problem in this instance would of course be magnified if the outbreak were widespread rather than localized and had only manifested itself in close proximity to the laboratory. In the case of the widespread outbreak policy factors might operate to encourage the courts to require strong evidence of causation. That result could easily be achieved by placing emphasis on the need to establish that the release of the virus or organism from either a factory or a laboratory was a sufficiently proximate cause.

Several tests of causation have been formulated by the courts, although the "but for" test is the one that is most commonly applied.[33] In this context it is not a particularly useful means by which to demonstrate that the relevant damage was caused by the negligent act in question. For example, if a genetically engineered virus were negligently released into the community, it could be argued that a particular plaintiff would not have become ill "but for" the release. Clearly, the test does not solve the evidentiary problems that are involved in showing that the illness was related to the virus.

Similar problems arise in relation to multiple and intervening causes. How could the courts deal with liability for the release of a virus that was known to lower resistance to infection in humans? In such a circumstance the claim for compensation would arise out of the subsequent infection rather than the viral infection because the latter would not, in itself, result in actual physical damage but only in an increased susceptibility to infection. Perhaps the plaintiff in this type of situation can be equated with the plaintiffs in *Beavis* v. *Apthorpe*[34] and *Smith* v. *Leech Brain*.[35] In the former case bacteria entered a wound that was caused by a car accident. Eighteen months later the plaintiff had a bone graft on his injured leg. This operation resulted in the onset of severe tetanus, which was attributed to the presence of bacteria that had lain latent since the time of the accident. In the latter case the plaintiff had a premalignant condition that developed into cancer when he received a burn. In both cases the plaintiffs were awarded damages, largely on the basis that the negligent acts that they complained of were thought to be sufficiently direct causes of their injuries. Although those cases are distinguishable from the genetic engineering example,* there are sufficient similarities on which to base an argument that the subsequent infection is not an intervening cause that breaks the notional chain of causation. On the other hand only one plaintiff was foreseeable in the *Smith* and *Beavis* situations. That cannot be said in the genetic engineering context, and accordingly the ubiquitous policy considerations that have come to

---

*This is particularly true of the *Smith* case in the sense that the negligent act was subsequent to the susceptibility.

play such an important part in the development of the tort of negligence may influence the courts against allowing plaintiffs to recover from negligent genetic engineers.

It is evident that many of the problems that have been discussed above in relation to genetic engineering also arise in other research contexts. Special tort rules have already been formulated for dealing with what have come to be recognized as special activities. In the case of *Rylands* v. *Fletcher*, Blackburn, J. held that a "person who for his own purposes brings on his lands and collects and keeps there anything likely to do mischief if it escapes must keep it in at his peril, and, if he does not do so, is prima facie answerable for all the damage which is the natural consequence of its escape."[36]

In such cases fault is not considered; the words "prima facie answerable" import a standard of strict liability as opposed to liability for negligence. Genetically engineered organisms appear to fit within this category.[37] They involve unusual activities that carry with them a high degree of risk to others. At the same time the usefulness of the technology and the fact that it can be argued that the risks are not intolerably large lead to the conclusion that those who carry out this activity should not be held to have been *negligent* simply because harm has resulted from their activities.[38] This accords with the rationale behind the rule in *Rylands* v. *Fletcher*, which divorces liability from negligence so that those who engage in useful but high-risk conduct will not be stigmatized, although they will be required to pay for the damage that is caused by their activity.

It should also be remembered that the rule in *Rylands* v. *Fletcher* was originally phrased in terms of escape.[39] However, in *Read* v. *Lyons*[40] the House of Lords narrowed the potential application of the rule by declaring that an escape, as contemplated in the *Rylands* case, must be an escape of a dangerous substance from land under the control of the defendant to land outside his occupation. Under the circumstances a laboratory technician or factory worker who sustains personal injury as a result of an accident within the laboratory would not be able to recover on the basis of the rule in *Rylands* v. *Fletcher*.

Fleming points out that the decision in *Read* v. *Lyons* has been applauded by some commentators as a welcome reminder that the rule in *Rylands* v. *Fletcher* is simply a branch of the tort of nuisance.[41] This is clearly correct in the sense that the only significant difference between the two causes of action is that the rule in *Rylands* v. *Fletcher* permits recovery for isolated escapes. The problems involved in proving causation are the same in both cases and it is now settled that proof of foreseeability is also an essential part of the nuisance action.[42]

The common law trespass action is also unsuited to claims that relate to damage caused by genetic engineering accidents. An alleged trespass must be direct rather than consequential and it would seem that infection by genetically engineered viruses or microorganisms would not be held to be so directly linked with the conduct complained of that it would constitute trespass to the person or property of the plaintiff.

Although existing torts could conceivably form a basis for the recovery of losses that are caused by genetic engineering accidents, the possibility of such recovery is speculative and uncertain. Does any existing statute offer greater hope of compensation? The Accident Compensation Act 1972 replaces, with a statutory remedy, common law actions relating to personal injury by accident.[43] Compensation is guaranteed irrespective of fault or negligence if the injury complained of can be said to constitute a personal injury by accident within the meaning of section 2 of the act. Damage to the body or mind caused exclusively by disease, infection, or the aging process is specifically excluded from the definition and these exclusions could be thought to preclude recovery for personal injury caused by genetically engineered viruses or microorganisms.[44]

Under the Accident Compensation Act a distinction must be made between a member of the public who sustains injury as a result of the use of genetic engineering techniques and a researcher who is injured at work. In the latter case the injured worker may be able to recover his losses in accordance with section 67 of the act. Under that section compensation is payable for diseases that arise out of employment, as if the disease were a personal injury by accident. Some of the common law problems that relate to proof of causation are still embodied in the section. For example, it must be apparent that the claimant's disease was "due to the nature of the employment" in which he was employed at the relevant time. Two decisions of the Accident Compensation Appeal Authority are of particular interest in this context. In *Rosson's* case[45] Blair, J. refused to compensate a woman who claimed that she had contracted dermatitis as a result of her employment as a cleaner, in spite of the fact that an eminent skin specialist had reported that the dermatitis "arose from her work." Blair, J. was of the opinion that this did not constitute "sufficient evidence to incriminate the *work process* itself"[46] and went on to hold that the disease was not due to the nature of the woman's employment.

The appellant in the *s* case[47] claimed that he had developed pneumonia with pleurisy as a result of working in the rain without waterproof clothing. Blair, J. referred to Lord Denning's judgment in the *Pyrah* case and noted that that case had yet to be argued in relation to the Accident Compensation Act. He then left this issue open by assuming, without expressly deciding the point, that "there may be circumstances in which the contracting of a disease will give entitlement to compensation under the Act."[48] However, in the end result the appellant failed because he was not able to establish a causal link between the drenching and the pneumonia.

In practical terms it should not be unduly difficult to establish a causal link between a disease and employment in a factory or a laboratory in which genetic engineering techniques are being used. It should be possible to detect evidence of the offending microorganisms in blood and tissue samples taken from the complainant for comparative purposes. Despite the fact that the chemical and genetic composition of the organisms can change even after release within an enclosed environment, proof of identity and origin should not be impossible.[49]

Section 67(1) provides that the claimant must have been engaged in the employment from which the disease is alleged to have arisen during a prescribed period before the date of the commencement of the incapacity. In the case of unspecified diseases that period is two years.[50] This unfortunate limitation should be abolished. It cannot be justified on economic grounds in view of the fact that in terms of the scheme as a whole its impact is likely to be negligible. If the relevant consideration is whether or not a casual link can be established, then there is an unjustifiably harsh presumption that claimants who have been employed for less than two years in the job in question did not contract their disease because of the nature of their employment. The length of time employed, if it is to have an influence at all, should operate only as a factor from which inferences about the probabilities of the relationship between the employment and the disease can be drawn—it should not in any circumstances be treated as a complete bar to recovery.

Section 67(8) is also of some interest in the context of genetic engineering. It provides that nothing in section 67 "shall affect the right of any person to recover compensation in respect of a disease if the disease is a personal injury by accident within the meaning of (the) Act." Those words may indicate that diseases other than those that arise from employment can constitute personal injuries and indeed this is closely related to the point that Blair, J. left open in the s case.[51] Apart from section 2, the only sections in the act that refer to disease are sections 65 and 67, both of which concern occupational disease. Thus section 67(8) tends to suggest that the exclusion in paragraph (b) (ii) of the definition of personal injury is not decisive. However, the word "exclusively" may provide the key in the sense that section 67(8) presumably envisages that damage to the body or mind caused partially by disease can constitute a personal injury by accident in relation to which compensation can be recovered.[52]

Although the Accident Compensation Act offers some assistance to workers who are injured in the course of their employment, it does not provide a salve to members of the public who suffer loss as a result of the new technology of genetic engineering. Accordingly, those whose diseases cannot be described as occupational may have to rely on the unsatisfactory common law remedies that will still be open to them if their injuries are said to be caused solely by disease or infection.

It seems clear that existing legal mechanisms, whether common law or statutory, do not offer satisfactory means of preventing genetic engineering accidents or compensating for the damage that they could cause. However, it is possible to envisage a more appropriate means of control in the form of a statute devoted solely to the technology of genetic engineering.

A NEW STATUTE

Aims and Scope

An act drafted specifically for the purpose of controlling genetic engineering should aim toward the establishment of a unified system of regulation that would, to the fullest extent possible, govern every facet of the technology. As a reflection of the importance that society attaches to this science, the act could begin with a broad statement of intent regarding the protection of animal and plant life, natural resources, and the environment.

The enactment of a statute that would govern individuals, companies, and governmental institutes is consistent with the concept of complete regulation and control. Accordingly, the act should apply to all genetic engineering experiments that are conducted and to all genetically engineered organisms that are created or used or imported into a country.

It is clear that an effective Genetic Engineering Act should extend to companies. In fact the current problems that are involved in controlling nongovernmental genetic engineering research by means of guidelines have, to a large extent, been responsible for the recognition of the need for legislation.[53] On the other hand it has been argued that the nonlegal controls that are exerted by peer group pressure and guidelines are sufficient to control the experiments that take place in universities and governmental institutes.[54] Such faith in public sector activities may be misplaced. In the United States peer group pressure has, on more than one occasion, failed to operate as an effective means of control over this research. In addition, a genetic engineering statute would, for example, deal with issues concerning compensation and emergency procedures. Those issues are equally applicable in corporate and noncorporate settings. It should also be remembered that, at present, genetic engineering research in New Zealand is the exclusive preserve of government-funded researchers, although that situation will change if pending patent applications are granted.*

---

*See applications 187300 and 163521, which are discussed in detail later in this chapter.

The U.S. experience in this area is instructive.* Many of the bills that have been introduced into the House and the Senate encompass the widest possible class of potential offender by defining the word "person" to include any individual, partnership, corporation, governmental entity, or group of individuals. Thus the legislators declare that the acts apply to all genetic engineering research that is conducted by any person in the country, although, in some instances, they have neglected to cover the practical industrial applications of the research. How could genetic engineering be defined for the purposes of the act?

Definitions

To an extent the question that is posed above has already been answered in Chapter 1 of this book. It is clear that any limitation that is placed on the meaning of genetic engineering is necessarily subjective in view of the wide range of activities that can be covered by the term. The Irvine Committee made every effort to avoid mentioning genetic engineering. It preferred instead to refer to novel genetic techniques.[55] Despite the fact that the phrase is by no means self-explanatory, it has some advantages. One factor that may have influenced the working party could have been the desire to avoid confusion between recombinant DNA techniques and genetic engineering techniques. Although genetic engineering covers a far wider field than recombinant DNA research, the two techniques have often been associated with each other in articles in magazines and newspapers.[56] In addition, many of the guidelines that have been formulated overseas apply only to recombinant DNA research. It has also been argued that it is easier to attach a pejorative meaning to genetic engineering techniques than to novel genetic techniques,[57] and once again this may be attributable not only to the connotations of the word "engineering" but also to popular usage.[58]

The working party recommended the establishment of a national committee for the purpose of controlling the use of novel genetic techniques in New Zealand. It proposed that the national committee adjudicate on novel genetic techniques, which include the following:

(a) any procedures involving the combination of DNA or RNA molecules of different biological origin by means that overcome natural barriers in mating and recombination, to yield molecules that can be propagated in some host cell, and the subsequent study of such molecules;

---

*See Chapter 3, first section.

(b) any procedures involving the combination of chemically or enzymically prepared copies of DNA or RNA molecules of different biological origin to yield molecules that can be propagated in some host cell, and the subsequent study of such molecules;

(c) with the specific exceptions given in (f), (g), and (h) below, the fusion of animal, plant, fungal, or bacterial cells inter-specifically by whatever means, leading to the formation of cells or complete organisms with novel genetic constitution;

and exclude the following:

(d) standard procedures for the uptake by eukaryotes *or* eukaryotic cells of homologous exogenous DNA or RNA or unmodified DNA or RNA derived from a virus which has been shown to infect that eukaryote or eukaryotic cell;

(e) standard procedures for the generation and selection of genetic variants;

(f) procedures involving animal cell fusion, whether mediated by virus or not, involving cells of different lineage and performed specifically for the purpose of study of the genetics of such cells;

(g) procedures involving the fusion of protoplasts of cells of related species of flowering plants;

(h) standard procedures of DNA transfer such as transduction, transformation, transfection, and mating employed for the study of the genetics of bacteria and their viruses and performed for that specific purpose.

Any proposal for experiments involving novel genetic techniques not specifically excluded by categories (e), (f), (g), and (h) must be referred to the Committee before any experiments are initiated. If the need arises, (a) to (h) would be changed at the discretion of the Committee.

By adopting this approach the Irvine Committee avoided the need to actually define what was meant by novel genetic engineering techniques in terms of stating, for example, that "a novel genetic technique is any technique that involves . . . ." Although the committee's method is not totally lacking in merit,* it is arguable that it leads to uncertainty. For example, it is not entirely clear that the experiments in categories (a) to (c) inclusive are novel genetic techniques and that those in categories (d) to (h) inclusive are not, or conversely that the experiments in categories (d) to (h) are novel

---

*The inclusion of RNA recombinations was forward-looking and desirable. The categories listed can also include *in vivo* research, although this does not appear to have been contemplated when the MRC/DSIR Guidelines and the DSIR Advisory Committee Report were drafted.

genetic techniques upon which the committee does not have the power to adjudicate. The latter interpretation would seem to be the more reasonable. It is supported by the requirement that proposals for "experiments involving novel genetic techniques not specifically excluded by categories (d), (f), (g), and (h) must be referred to the National Committee before work can begin. However, even on the basis of that interpretation the list of categories is unsatisfactorily large because the working party purports to deal with definitional and judicial issues in the same paragraph. The consequence of this is that it will not be clear which projects have to be submitted to the committee. For instance, what is the position when an experiment does not appear to fall within any of the categories of inclusion and exclusion? The last paragraph of the report suggests that if such an experiment involves novel genetic techniques then it must be forwarded to the committee, but who is to decide whether a particular technique constitutes a novel genetic technique? The working party has not, as such, defined that term. Those involved do not have an obligation to submit proposals that do not relate to novel genetic techniques and for that reason the committee may never have the opportunity to adjudicate on some important projects.

This defect may be partially ameliorated by the ease with which the committee could alter the categories if it were empowered with the recommended discretion to do so, and in such a highly technical field it would seem reasonable to vest the power to make such decisions in the hands of a specially constituted committee. Presumably, high-risk activities that are not presently listed in the proposed categories of inclusion will be added as the need arises, but nevertheless it would be desirable to draft a relatively wide definition of a novel genetic technique in order to ensure that the researchers involved do not usurp or exclude the decision-making powers of the committee.* It can be said that the inclusion of a wide definition would place an unnecessary strain on the resources of the proposed committee in the sense that it would be deluged with experiments of the type that are currently being conducted by university students and, more exceptionally, by secondary school students. That argument can be answered in two ways. First, experiments and applications that are thought to involve small or nonexistent risks can be excluded by appropriate exclusionary categories. Second, if a statute is to be passed and a special system of regulation, administration, and enforcement established, then the bodies that are charged with those functions should be adequately equipped to review relatively large numbers of proposals.

Indeed the Irvine Committee suggested that if an act were passed it should extend to the regulation of a category of "microbiological hazard."[59]

The categories could still be retained by way of example.

This approach was rejected by those who formulated genetic manipulation regulations in Britain. The Genetic Manipulation Advisory Group stated:

> In considering the definition of genetic manipulation appropriate to our own functions, we accepted, with the Williams and Ashby Working Parties, that, insofar as conventional techniques, used for many years with no evidence of hazard, existed to attain apparently the same ends as the new ones, the outcome was reassuring. It is only the possibility of some special untoward extra hazard that' justified special control for the new techniques. It was therefore judged inappropriate to bring under control the conventional techniques.[60]

These reactions hardly seem sufficient to justify the Advisory Group's conclusion. The assertion that conventional techniques have been used for many years without evidence of hazard is particularly surprising. In the British context the smallpox accidents, the foot and mouth outbreak that was the subject of the *Weller* decision, and similar incidents at Porton Down serve as stark reminders of the dangers that are inherent in microbiological research. The fact that they involve dangers that are supposedly well known does not detract from the desirability of attempting to control them.

While it is clear that an accident resulting from the use of a novel genetic technique could prove to be more catastrophic than an accident occurring during the use of a conventional technique, the probability of the occurrence of the event in the latter case is higher than in the former. Conventional techniques are in widespread use in schools, hospitals, universities, governmental institutes, and company laboratories. The DSIR Advisory Committee pointed out that between 200 and 300 people are presently engaged in microbiological research in New Zealand and noted that "it might well be asked why a select few genetic engineers should be trusted less than other scientists."[61] The British views on microbiological hazard do not provide a satisfactory answer to that question.

Detailed consideration of that issue is outside the scope of this book, but it can be argued that, in practical terms, microbiological hazards should be controlled under the proposed new act. Conventional techniques and novel genetic techniques could be controlled by the same committee or committees, by the same inspectors, and the same sanctions. In January 1973 the N.Z. Department of Health published a set of guidelines that apply to laboratory safety in microbiological laboratories.[62] The precautions listed are very similar to those that relate to category I facilities as described in the MRC guidelines. Accordingly, the Department of Health's recommended safety precautions, although they are out of date and require modification, could, in revised form, be implemented as part of a set of regulations that could be promulgated under the proposed new act.

Committee Structures and Functions

The proposed act would provide for the establishment of a three-tiered committee structure. At the local level an Employer Safety Committee would be appointed. In addition, a National Advisory Committee would be established and the top position in the hierarchy reserved for a Supervisory Commission.*

A Proposed National Supervisory Commission

A National Supervisory Commission with a wide range of functions could draw its members from a number of diverse sources. To ensure a wide spread of interests and occupations, the ministers of Health and Science and Technology should appoint 12 members to sit on the commission for a period of three years. This independent body would be composed of six scientists (comprising not more than three from government, one from private industry, and two specialists in the field of novel genetic techniques), one trade unionist, an industrial representative, one specialist in sociology or an ethics-related field, one lawyer, and two lay members. The commission would be empowered to appoint consultants to advise it on any issues relevant to its duties. Its primary duty would be to advise the government on the promulgation of a set of statutory regulations† for the purpose of establishing a mandatory code of practice for the conduct of novel genetic techniques. The act should provide for the promulgation of regulations that are at least as stringent as the modified MRC guidelines. Beyond that it would, for example, specify that the regulations could ban certain experiments, specify the appropriate containment levels for others, provide for the establishment of health monitoring programs for employees, and set out procedures for the transportation of microorganisms.[63] Such regulations could be frequently and easily amended in order to maintain the degree of flexibility that is required for the control of fast-developing techniques.‡

In view of the need for periodic review of the regulations, the second major function of the Supervisory Commission would be to study the implications of genetic engineering research. This would involve an evaluation of the risks and benefits that are attached to techniques currently in use in the country and a review of new scientific and safety developments over-

---

*A three-tiered structure was suggested by the MRC Advisory Committee but ignored by the Irvine Committee in its working party report.

†Compare the role of the Securities Commission under sections 10 and 70 of the Securities Act 1978 (N.Z.).

‡In practice the Supervisory Commission would rely, to a large extent, on the advice of the National Advisory Committee, which would be able to offer considerable technical assistance in the relevant areas.

seas.* In this context the Supervisory Commission could cooperate with the Environmental Council, which is an independent body that acts in an advisory capacity to the government on matters of environmental policy.[64] It is responsible for the publication of such information as it considers necessary in the public interest and for advising agencies such as the Commission for the Future and the Planning Council.

It is also arguable that the Supervisory Commission should be empowered to conduct public hearings for the purpose of gathering information and gauging public opinion on certain types of experiments and procedures. The commission could be required to present an annual report to Parliament that would include a description of the state of the technology in New Zealand, a prediction of trends, and a summary of any action that has been taken by the commission in terms of drafting or proposing amendments to regulations and establishing safety training programs for those who work with novel genetic techniques.

A Proposed National Advisory Committee

It is possible to envisage a National Advisory Committee that would ideally be comprised of six scientists, two of whom would be specialists in the field of novel genetic techniques.† The other four should have a more general biological background and could include one government scientist, one university researcher, a member of the Medical Research Council, and a private sector scientist.

The most important function of this committee would be to administer the licensing provisions that would form an essential part of the new act. Before any "person," as defined in the act, could commence work on Category II, III, and IV‡ experiments in New Zealand, a license for the project in question would have to be obtained from the director-general of health. The licensing provisions would thus apply equally to private and public sector research.[65] The committee would consider the research proposals submitted to it by project supervisors,** via their Employer Safety

---

*For this reason it would be desirable to allocate funds for the purpose of liaison with the relevant organizations in other countries.

†Provided that it would not be possible for any person to sit on both the National Supervisory Commission and the National Advisory Committee.

‡These categories are described in the appendix to the MRC Guidelines, op. cit. Researchers would also have to notify the committee about Category I experiments, although the licensing provisions would not apply to them.

**The Employer Safety Committees would have an obligation to declare the names of project supervisors on the relevant license application forms. The licenses would be granted for a one-year period. At the end of that time an application for renewal could be filed.

Committees. It would then advise the director-general of health as to whether he should grant a license to the project supervisor in respect of the proposed activity.* If the committee was of the opinion that a particular project was unacceptable either in terms of the unsuitability of the host facility† or a failure to comply with the regulations and the director-general accepted their opinion, the committee would return the proposal to the project supervisor, accompanied by recommendations and an explanation of the reasons for the refusal of the license. The act should provide for a maximum period of six weeks between the date of notification by the project supervisor and the granting of a license or the return of a proposal with recommendations attached.⁶⁶

The licenses would be granted subject to certain specified limitations and conditions that could be varied at the discretion of the director-general. In the case of any breach of a license,‡ the director-general should be given a discretion to cancel or suspend the license for such period as he thinks fit, subject to certain specified rights of appeal. A breach of a license by a person acting under the instructions or supervision of the applicant should be treated as a breach by the applicant unless he can show that the breach was committed without his knowledge and that he had exercised all due diligence in attempting to comply with the conditions of the license.** In addition, the director-general should have the power to revoke or vary a license at any time, simply because he considers that it is in the public interest to do so.††

The National Advisory Committee would be empowered to inspect factories and laboratories and, as the working party recommended, it would be appropriate for at least one committee member to visit the site of each proposed activity.⁶⁷ Hence the committee's primary function in this context would be in terms of project approval rather than enforcement. The Advisory Committee should also be responsible for the establishment of a

---

*Alternatively, it would be possible to vest the power to grant and revoke licenses in the Advisory Committee. See Section 3 of the Agricultural Chemicals Amendment Act 1970. However, compare sections 16 to 22 of the Radiation Protection Act 1965. It can be argued that decisions that may have a substantial effect on the health of the general public should not be left in the hands of a statutory board or committee.

†At present New Zealand has no Category III or IV containment-level laboratories. Accordingly, it may be necessary to modify one of the existing Auckland laboratories or to conduct high-risk experiments overseas.

‡Offenses and penalties under the new act are discussed in the last section of this chapter.

**Compare Radiation Protection Act 1965, section 27, which relates to the licensee's liability for offenses committed by others.

††Such a provision is not without precedent. See the Radiation Protection Act 1965, section 20.

national register of the projects, project supervisors, and institutions that are involved in genetic engineering research in New Zealand. Such registers are already being compiled in other countries and they appear to represent a desirable development not only in terms of control but also from the point of view of increased public awareness.

On the other hand it can be argued that the number of researchers engaged in genetic engineering experimentation in New Zealand is not sufficient to justify the existence of a national register. While that proposition can be disputed,* the idea begins to lose some of its superficial attractiveness when it is considered in light of the problems raised by the need for confidentiality in a competitive setting. In that context it is evident that, at least in terms of the details of the projects involved, the public's right of access to the relevant information would have to be a limited one. It has already been proposed that provision should be made for the certification of the vectors and hosts that are used in genetic engineering experiments.[68] On this issue the recommendation of the working party is eminently practical and should be adopted. The working party suggested that the responsibility for certification should rest with the National Institute of the Department of Health, acting on the advice of a committee that would be similar to the National Advisory Committee that is proposed in this book.[69]

Finally, the National Advisory Committee should be required to publish an annual report on its activities. This report would be similar to the "First Report of the Genetic Manipulation Advisory Group"[70] and would include a description of the type and number of proposals submitted, the time taken for the consideration of proposals and license applications, and a discussion of any problems that were encountered in administering the act and regulations.

A Proposed Employer Safety Committee

A three-member Employer Safety Committee headed by a biological safety officer would provide much needed control at the local level. All three members of the committee should belong to the institution that is using novel genetic techniques and, if possible, one person should be a

---

*The Irvine Committee stated that in New Zealand "fewer than ten" people were conducting experiments that involved novel genetic techniques. However, that figure appears to have been based on a survey that was conducted by the DSIR Advisory Committee on Novel Genetic Techniques in October 1977. In its report the Advisory Committee, after describing the work that was being carried out in a number of laboratories, stated that "it is not claimed that this summary is exhaustive. The main institutions not canvassed were the Ministry of Agriculture and Fisheries and almost the entire Medical Research establishment (hospitals, Medical Schools and National Health Institute). Minor projects in the DSIR may have been overlooked. Finally no attempt has been made to survey work with plant and animal viruses."

lay representative. For example, in the case of research conducted by a university, a lay member of the University Council could be co-opted.[71]

This committee would, in the first instance, review activities proposed by individuals within the organization. On the basis of a full statement prepared by the project supervisors it would determine whether the proposal should be submitted to the National Advisory Committee. In that sense it would be responsible for deciding when a proposed project falls within the ambit of the act. It can be predicted that in many cases such a committee would veto proposals that fail to meet the requirements that are specified in the regulations. In this way it would alleviate some of the pressure on the National Advisory Committee. In fact the committee would have a vested interest in ensuring that only the most carefully planned applications were forwarded to the Advisory Committee. If an ill-considered proposal were submitted, the Advisory Committee would be likely to advise the director-general to grant a restricted license or to refuse a license altogether. In the former case the institution would have to review its project and resubmit it. In addition to the fact that that process is time consuming, an initial rejection by the Advisory Committee could easily prejudice a subsequent application.

The MRC Guidelines contained a recommendation to the effect that a biological safety officer should be responsible for informing staff members of the nature of the work in which their colleagues propose to engage.[72] This suggestion is a useful one that is presumably aimed at increasing the efficiency of informal peer group pressure controls. The safety officer would also be responsible for ensuring that the project supervisor keeps proper records. The other major functions of the Employer Safety Committee should include the establishment of a health monitoring program for employees,* the implementation of localized training programs in accordance with the directives of the National Supervisory Commission, and the preparation of a short statement of activities for the information of the National Advisory Committee.

Inspection

In order to ascertain which agency is in the best position to provide suitable inspectors to assist in the enforcement of the provisions of the proposed statute, it is necessary to examine the possible duties of the inspectors and the powers that would be vested in them under the act.

---

*This could, for example, involve annual serological checks for the presence of antibodies to pathogens that are in use within the organization.

On the presentation of documentary authority and on the grounds of a reasonable belief that novel genetic techniques are being used, the inspectors should be able to enter any premises for the purpose of investigating the manner in which the activity in question is being conducted. Samples could be removed in appropriate cases and there would be an investigation of the adequacy of the relevant facilities. In each case the inspector would prepare a written report of his findings and forward it, within a reasonable time, to the Employer Safety Committee of the institution concerned. Some of the U.S. recombinant DNA bills place a more immediate duty on the inspector (see Chapter 3). For example, "When an inspector has completed such an inspection he shall, *before he leaves the facility* [emphasis added] inform the owner, operator or agent in charge of the facility of any conditions or practices which in the inspector's judgment constitute a violation of any of the requirements of this Act."[73] The question that arises in this context is whether an independent agency with specially trained inspectors should be set up specifically for the purpose of enforcing the provisions of the statute. It is clear that the proposed National Supervisory Commission could not act as a suitable inspectorate, and although the six members of the National Advisory Committee would have the relevant expertise, they would be fully occupied with their administrative duties and site visits in connection with licensing decisions. On the other hand the establishment of a separate, new inspectorate should be avoided if the relevant functions can be adequately discharged by existing agencies or government departments.

The Commission for the Environment lacks the power that participation in the enforcement process requires. Despite frequent calls for legislation that would give the commission statutory independence,[74] such legislation has not been forthcoming.[75] Local authority inspectors are already overburdened and quite apart from that factor it is clear that uniform national control is preferable to local body enforcement, which is likely to be differential. Local inspectorates can also be argued to be particularly inappropriate in countries like New Zealand in the sense that relatively few institutions are involved in this research.

Similarly, an inspectorate drawn from the staff of the Department of Scientific and Industrial Research would not be satisfactory. The primary function of the DSIR is "to initiate, plan and implement research calculated to promote the national interest of New Zealand."[76] To request the DSIR to inspect would be almost as reprehensible as placing control in the hands of the National Research Advisory Council, which has a strong vested interest in "the promotion and development of scientific research in New Zealand."[77]

There appears to be no reason why the Health Department could not supply inspectors. The department provides an almost identical service under the Radiation Protection Act 1965.[78] In fact it may not even be necessary to appoint a large number of new officers. Inspectors could be seconded from the Bureau of Public Health and Environmental Protection,[79] the Departmental Committee on Pollution of the Environment, or perhaps most appropriately, from the Occupational Health and Toxicology Branch.

Emergency Procedures, Injunctions, and Applications in the Nature of Mandamus and Prohibition

Provision must be made for statutory powers that could be exercised if a dangerous situation manifested itself in the community or was discovered in the course of an inspection. In other words legislation that purports to preserve the environment must contain provisions that support legal controls with the possibility of effective containment and eradication. The procedures involved should be designed for swift and decisive implementation in the event of an accident. They would have statutory analogues in section 30 of the Animals Act 1967 and section 12 of the Plants Act 1970, which relate to the proclamation by the governor-general of states of animal and plant disease emergency. During an emergency proclaimed under those acts the minister of Agriculture has extensive powers at his disposal. He can, for example, require registered veterinary surgeons who reside anywhere in New Zealand or any fit male person over the age of 18 years who resides or works within a five-mile radius of the place where his help is required to assist in the eradication or limitation of the spread of disease.[80]

Similarly, under section 30(4) (b) of the Animals Act any person who owns any piece of equipment that could be used to control an outbreak of disease can be required to transfer it to the minister for the period of the emergency. Those sections are supported by section 31 of the act, which is a very broad empowering provision.[81] It stipulates that if a state of emergency has been declared under the act the minister may, in the relevant area or areas, "take such measures, and do all such acts and things, and give all such directions, and require all such acts to be done, as in the opinion of the Minister are necessary and desirable for the purpose of eradicating the disease or preventing or limiting the spread of the disease."

Section 33 of the Dangerous Goods Act 1974 is also of interest in this context. It provides that if an accident involving dangerous goods occurs on any premises that are licensed under the act the licensee must send notice of the accident to an inspector having jurisdiction over the relevant premises. The notice must contain details of any death, personal injury, or property damage that is occasioned by the accident.[82]

This is a very useful provision that might well be included in proposed genetic engineering legislation in the sense that it would facilitate investigations into genetic engineering accidents. New Zealand's record in this matter has not been beyond reproach in the past. When the accident that destroyed the pine seedlings occurred at Palmerston North (see Introduction), the diseased plants were autoclaved before there was an opportunity for an inquest.[83] Under the circumstances the action that was taken can arguably be justified on the basis that the primary consideration in the case of an accident is the prevention of the spread of damage, and indeed it cannot be denied that in some cases inquiries that might aid the prevention of similar incidents should give way to the exigencies of the situation. Nevertheless a certain amount of investigation is both desirable and consistent with containment.

Provisions of the type that have been described above would form an essential part of legislation for the control of the technology of genetic engineering. Seizure, treatment, and destruction procedures would all have an important place but, like all administrative powers, they must be exercised with restraint. It will clearly be necessary to empower the minister of health to seize or take control of dangerous laboratory pathogens even if harm to the environment is thought to be imminent rather than actual. In all these cases the minister should not be empowered to act other than in accordance with certain prescribed standards. For example, he could be required to take action if seizure is considered necessary to preserve the public health and safety.*

If the minister fails to act, when action is arguably required in the public interest, concerned individuals should be able to institute proceedings for judicial review in the nature of mandamus. The minister would be compelled to perform his public duty under the act if the application were granted. Conversely, an order in the nature of prohibition could be sought by a person who claims to be affected by certain proposed actions of the minister.† However in this area, as in many others, the requirement that an individual who brings such an action must have the standing to do so could present a practical bar, not only in relation to the writ but also in a wider context.[84] Before considering the *locus standi* requirement it is necessary to look beyond the restricted area of applications for judicial review. Although applications for relief in the nature of mandamus and prohibition can be viewed as reasonably appropriate means of controlling administrative action, there is another remedy that can be used not only for the purpose of controlling administrative agencies such as the proposed National Advisory Committee, but also for the purpose of controlling those who conduct genetic engineering experiments and use genetic engineering

---

*Several New Zealand statutes already contain provisions to that effect. For instance, section 34 of the Dangerous Goods Act 1974 refers "to any incident which creates a threat of explosion or fire, or creates a hazard which threatens the public safety . . . ."

†Certiorari is discussed in the next section, which is devoted to the question of appeals against administrative decisions.

techniques. That remedy is the equitable injunction. Proceedings could be instituted either by the minister of Health on the advice of the Supervisory Commission or the National Advisory Committee, or by a member of the public. The target is also variable in the sense that the courts could grant an injunction against companies, private individuals, and governmental research institutes like the DSIR.*

The proposed statute should contain provisions that relate specifically to injunctions.[85] The remedy could be sought through the courts in the normal way, but a standard by which to gauge a proposed or continuing activity should be written into the act. For example, it could be stated that either the minister or any person who has reason to believe that the continuation of certain activities constitutes *a significant danger to the public health or is not in the public interest* should seek a temporary or final injunction prohibiting the experiments in question.[86] Because the injunction is an important remedy it should be made available to *any person* who has reasonable grounds for believing that the continuation of particular genetic engineering activities would constitute a significant danger to the public health or interest. The "any person" usage clearly represents a substantial statutory liberalization of the usual standing requirements;[87] however, such a provision is justified not only because of the effectiveness of the injunction as a mechanism of control in this type of situation but also because of the sorry record of courts that have had to determine whether or not applicants for injunctions, declaratory judgments, and the prerogative writs have standing.

The *Environmental Defence Society Inc.* v. *Agricultural Chemicals Board* case illustrates some of these difficulties in the context of the writ of mandamus.[88] In referring to the Environmental Defence Society Haslam, J. expressed the opinion that

> while the dedication and sincerity of its members were not in dispute, it was contended that an incorporated body with objects so defined, could suffer neither physical detriment nor economic loss from the effects of 2,4,5-T, and that in essence its "special interest" was no more than the personal concern that might be felt by any responsible private citizen with strong views upon the dangers of chemical contamination.†

Responsible personal concern is clearly not sufficiently influential at common law and despite an encouraging, although recently reversed, trend,[89]

---

*In appropriate cases the courts could make declaratory judgments or orders under sections 4 and 7 of the Judicature Amendment Act of 1972. The position as regards rule 473 of the Code of Civil Procedure is somewhat less clear.

† The standing problem did not arise in the *Mundy* case because a relator action was brought.

statutory inroads into traditional judicial attitudes to standing are to be preferred.

In practical terms two factors may play a particularly influential role in deterring plaintiffs from seeking injunctions. The effects of both can be mitigated by statute. The first is the cost that is involved.* In the *Mundy* case the plaintiffs were not awarded costs in spite of the fact that the interim injunction was continued. This was on the basis that "the defendants acted innocently and showed commendable respect for this Court in ceasing all operations immediately after they heard of this application." Unfortunately, Haslam did not allow a similar concession for the "dedication and sincerity" of the members of the Environmental Defence Society in the *Agricultural Chemicals Board* case.[90]

The Recombinant DNA Standards Bill 1977 (U.S.)[91] contains an attractive solution to the problem of prohibitive cost. Clause 10(b) (2) provides that, notwithstanding any other provisions of law, any party in a civil action shall be entitled to recover reasonable lawyers' fees, the fees and costs of experts, and other reasonable costs of litigation if the court determines that the action served an important public purpose. Such a provision could well be included in the proposed New Zealand legislation.

The second inhibiting factor, particularly in the context of the injunction, is that of the precarious position of employees who seek to restrain activities that are conducted in their own research institutes. Their interests must be protected, especially in view of the fact that they are clearly in the best position to detect potentially hazardous activities before they manifest themselves in damage to the community. There is another proposed U.S. provision that could be used as a precedent in this country. Clause 1796.9 of the California Biological Research Bill[92] provides that "No research facility or other employer shall discharge any employee or otherwise discriminate against any employee with respect to his compensation, terms, conditions, or privileges of employment because the employee has commenced or caused to be commenced, or is about to commence or cause to be commenced a proceeding" in respect of activities conducted by or under the direction of the employer. Similar provisions apply to employees who testify, assist, or participate in such proceedings.

Appeals against Administrative Decisions

The first question that arises in the context of appeals against administrative decisions is whether a special body should be appointed under

---

*In the environmental context this is usually increased by the need to produce a considerable amount of technical expert evidence.

the proposed legislation for the purpose of hearing appeals in relation to the director-general's licensing decisions.[93] Clearly, the National Advisory Committee would not be an appropriate appellate body. One of its primary functions would be to advise the director-general on the granting and revocation of licenses, thus, in practical terms, it will usually be the decision maker. The National Supervisory Commission could be used as an expert appeal board. Its members would be sufficiently well qualified to adjudicate on the merits of the relevant decisions, but as Keith points out,

> all review of the merits of a decision is prima facie wasteful. The taking of effect of the original decision is usually delayed, and the time of possibly highly qualified people is taken up going through the same material and reaching a second decision. Positive reasons for this waste, for taking a second or third opinion, must be adduced. The major reason justifying a further opinion is that it is likely to be better; that it is likely to be as good is no reason for seeking it.[94]

The logic of that argument cannot be disputed and it would indeed seem wasteful to provide for an "on the merits" appeal to the Supervisory Commission, particularly in view of the need for a uniform administrative policy in the genetic engineering context. The same arguments apply to substantive review by the courts, but in the case of judicial review there is an additional inhibiting factor in terms of a lack of expertise. The courts have been only too willing to acknowledge this deficiency in the past and in fact they have arguably shirked a considerable proportion of their responsibility for review. The 2,4,5-T case provides an interesting example of judicial reluctance to review expert decisions in the context of proceedings for a writ of mandamus. In that case Haslam, J. noted that

> the board is, by its composition, endowed with a high degree of expert knowledge and prima facie has the requisite collective capacity to assess scientific data and findings. . . . if the board decided in sifting the mass of material before it, that Dr. Becroft's clear cut views should be given considerable weight, this court, in its discretion, should dismiss the application for mandamus which is designed to interfere with that result.[95]

While it is clear that the courts are not well-equipped to make decisions that concern complex scientific data, it is equally clear that a new act should make provision for appeals on questions of law. The courts are obviously the most appropriate tribunal to handle legal issues, but the problem of which court and which judges are best placed to fulfill this role is considerably more difficult. In the past the legislature has varied its approach to this question in an apparent attempt to accommodate the differing technologies that it seeks to regulate. For instance, section 21 of the Agricultural Chemicals Act 1959 establishes an appeal authority that has jurisdiction over complaints regarding the refusal or revocation of applications for the registration of agricultural chemicals. The appeal authority consists of a magistrate and two assessors, of whom one is to be appointed by the board and one by the appellant. Its decision is final.[96]

Sections 33 and 41 of the Clean Air Act 1972 are at the other end of the spectrum. They appear to allow appeals to both the Supreme Court and the Court of Appeal on questions of *fact* and law in the event of dissatisfaction with the decisions of local licensing authorities.[97] At the Supreme Court level the appeals are heard by the Administrative Division, which sits with two additional members who are not judges of the Supreme Court and whose names are drawn from a special list of additional members that is prepared by the secretary for Justice with the approval of the ministers of Justice and Health.[98]

An intermediate approach is evident in sections 25 to 30 of the Plant Varieties Act 1973. A three-member appeal authority is established and provision is made for a right of appeal to the Administrative Division of the Supreme Court on points of law alone. However, as has already been indicated, it is arguable that even that limited right of appeal is too extensive in the genetic engineering context. The need for administrative efficiency may outweigh the need for "a second chance," especially if the appellate body is similar in nature to the original tribunal and will not bring special expertise to the decision-making process. On the other hand legal expertise can and should be provided by the courts in terms of a right of appeal to the Administrative Division of the Supreme Court, solely on questions of law that relate to the refusal, revocation, or variation of genetic engineering licenses. The appeal should be lodged within one month of receipt of notice of the decision at first instance. The ruling of the Supreme Court should be subject to review by the Court of Appeal in the normal way, although once again there may be a countervailing interest in terms of the uniform and consistent development of the law relating to licensing decisions in this field. That interest and the need for early determination of these matters may also justify a statutory prohibition on applications for review that are not specifically authorized by the appeal provisions of the proposed act.[99]

Administrative decisions that are made by the director-general of Health on the advice of the National Advisory Committee could conceivably be reviewed by the ombudsmen if the National Advisory Committee were added to the list of departments and organizations contained in the first schedule of the Ombudsmen Act 1975, although resort to the ombudsmen might be precluded by the existence of a right of appeal elsewhere.[100] For instance, an application for review under section 4 of the Judicature Amendment Act 1974 could be lodged in relation to a licensing decision if, in terms of the old common law writ of certiorari, there had been a breach of natural justice or an excess of jurisdiction or if an error of law had been made on the face of the record.[101]

Freedom of Information, Confidentiality, and Patents

Employee vigilance and a statutory inspectorate should not be relied upon as the only means of enforcing the proposed genetic engineering legislation. The dangers that are inherent in decision making by insular technocracies have already been examined in earlier chapters. The mechanics of obtaining full public participation in the decision-making and enforcement processes remain to be considered.

The public cannot participate in the control of genetic engineering unless it has access to data concerning this research and its applications. At present members of the community have very limited access to such information.[102] Their rights are particularly restricted in the private sector and in that context information that is received from concerned organizational scientists constitutes one of the only means by which the public can gain insights into the activities of private institutions.

Even that restricted source is largely unavailable in relation to the public sector, in the sense that scientists who are employed by the government and who wish to release information or comment on governmental projects are constrained by the provisions of the Official Secrets Act 1951 and the Public Service Regulations 1964.[103]

The need for a free flow of information between state and citizen in a democratic society is clearly not limited to the dissemination of scientific knowledge. Broad-ranging reforms in the shape of freedom of information legislation have already been introduced in the United States and Sweden.[104] Arguably, the problems that have been encountered with those acts can be attributed to the fact that the statutes represent overly comprehensive attempts to balance the competing interests of freedom of information and confidentiality in several different areas. As Keith suggests, that process is one that might best be considered in particular contexts.*

Although the release of the details that are concerned would not be likely to constitute a threat to national security or foreign relations, two categories that have been jealously guarded in the past, there has been a traditional reluctance to disclose information that could affect proprietary rights. Scientific and environmental data often fall within that category.[105]

---

*Nevertheless it is interesting to note that the breach of the NIH guidelines that occurred at Harvard was uncovered as a result of a request for information under the Freedom of Information Act 1966 (U.S.).

However, a common justification for the refusal to release such information is pitched at a lower level. For instance, the minister of science and technology has stated that

> scientists must also consider very carefully whether it really is in the public interest to release particular information. I am sure that you appreciate that there are occasions when premature disclosures, before sufficient facts are known and while research is still being done to uncover them, can often be quite misleading and confusing to the public, which does not always have sufficient training or knowledge to evaluate information in the same way as a professional scientist.[106]

If the minister's basic assumption about lack of information is accepted, then at least a partial remedy lies in his own hands.

The relevant competing interests in both the public and private sectors have already been weighed by the legislature in the context of environmental statutes. Section 47 of the Clean Air Act 1972, which relates to the unjustified disclosure of information, is particularly interesting. After excluding provisions that relate to a duty to furnish information to persons who are authorized under the act to receive it, the provision continues:

> (1) Any person who discloses any information relating to any manufacturing process or trade secret used in carrying on or operating any particular undertaking or equipment or plant which has been furnished to him or obtained by him under this Act or in connection with the execution thereof, commits an offence unless the disclosure is made
>
> (a) With the consent of the person carrying on or operating the undertaking, equipment or plant; or
> (b) In connection with the execution of this Act; or
> (c) With the prior permission of the Minister; or
> (d) For the purposes of any legal proceedings arising out of this Act or of the report of any such proceedings.
>
> (2) Without prejudice to subsection (1) of this section, no person shall be compelled to divulge in any civil proceedings any information that has been furnished to him or obtained by him under this Act.[107]

Similarly, section 26(1) (c) of the Radiation Protection Act 1965 provides that "every person commits an offence against this Act who . . . discloses any information obtained by means of the exercise of any power under section 24 of this Act." Paragraph (d) of section 47(1) of the Clean Air Act may not constitute as wide an exception as would first appear to be the case, in the sense that the courts have, in the past, demonstrated considerable reluctance to order the disclosure of either governmental[108] or private sector information.[109] In the *Distillers* case Talbot, J. placed primary emphasis on the need for confidentiality.[110] He noted that

> the public have a great interest in the thalidomide story (and it is a matter of public interest), and any light which can be thrown on to this matter to obviate any such thing happening again is welcome, nevertheless the defendants have not persuaded me that such use as they proposed to make of the documents which they possess is of greater advantage to the public than the public's interest in the need for proper administration of justice, to protect the confidentiality of discovery of documents. I would go further and say that I doubt very much whether there is sufficient in the use which the defendants have proposed to raise a public interest which overcomes the plaintiff's private right to the confidentiality of their documents.

Nonetheless, section 47 could provide a useful precedent for an "unjustified disclosure" section in proposed genetic engineering legislation,[111] although it can be argued that paragraph (c), which relates to disclosure with the prior permission of the minister, is too broad. Accordingly, it should be limited by the addition of the words "under circumstances which necessitate such disclosure in the public interest."

Proprietary rights achieve their highest form of expression in the form of a patent. The patent establishes such exclusive rights that the need for confidentiality disappears. On the other hand, the crucial period in relation to disclosure may precede the approval of a patent application. This may be one reason for delaying the publication of the details of projects that are the subject of applications for genetic engineering licenses.* The relevant descriptions or applications could be published with impunity after patents have been granted.

Can genetically engineered organisms be patented? Does the Patents Act 1953 restrict the grant of a patent to processes in which genetically engineered organisms are used, or can a patent be granted in relation to an organism? These questions are not purely academic. The New Zealand Patent Office has already received two applications from U.S. companies that are seeking patents over genetically engineered organisms.† There have also

---

*On the other hand, the criteria upon which license applications are assessed should be published by the Advisory Committee.

†No. 187300 concerning "Recombinant DNA Transfer Vectors and Micro-organisms containing a Gene from a Higher Organism" was filed on May 17, 1978, by the Regents of the University of California. Ranks, Hovis, and McDougall Limited filed application No. 16321 in relation to "Improvements In Or Relating to Micro-organisms" on April 30, 1971. Both applications are pending. There may be other relevant applications that are not readily identifiable from their designations.

been several applications overseas, one of which has been successful.*
Nevertheless it is likely that the New Zealand Patent Office will not only
refuse applications in respect to microorganisms but also applications concerning processes that involve the use of microorganisms.†

Under the Patents Act a patent can be granted in relation to an invention, which, as defined in section 2 of the act, means "any manner of new
manufacture . . . ." The definition also includes a reference to the Statute
of Monopolies 1628 (U.K.), which employed a "manner of new manufacture" test. As Gault points out, this reference effectively incorporates, into
New Zealand law, the English cases surrounding the interpretation of a
manner of new manufacture. Hence the word invention can be said to include new processes as well as new end-products.[112] In practice, however,
the English attitude to what is meant by a manner of new manufacture has
tended to be less restrictive than that which prevails in New Zealand.

The highwater mark in New Zealand was reached in *Swift and Company v. Commissioner of Patents*.[113] In that case the Supreme Court held
that the fact that the process for which a patent is claimed is a biological or
physiological invention is not a bar to the grant of a patent. The English
courts have approved the grant of patents for cosmetic techniques[114] and
processes, such as brewing, which involve the use of microorganisms.[115] On
May 19, 1976, the British Patent Commissioner granted the world's first patent over a genetically engineered organism and certain processes relating to
its production and use. The patent applies to a bacterium from the genus
Pseudomonas, which has oil-consuming capabilities.

Although a patent can be challenged on the basis that it is either not
new or that it is obvious,[116] the applicants acknowledged that existing conventional strains of *Pseudomonas* can decompose particular crude oil components.[117] Thus, even before the new variety of *Pseudomonas* was invented, oil spills could be biologically controlled by the use of a mixture of
bacterial strains that were each acknowledged to be capable of degrading
only a single component of the oil. However, this was unsatisfactory
because the various strains that were employed in this process had differing
nutritional requirements and rates of growth and the bulk of the oil often remained unattacked for several weeks, during which time it could spread or
sink. Accordingly, it could be claimed that the strain that has been
developed by Chakrabarty is truly "new" in the sense that it has a unique
genetic composition that enables it to function in a novel fashion.

---

*Patent specification 1,436,573 (U.K.) was granted on May 19, 1976, to the General Elec-
tric Company in relation to microorganisms and processes invented by Ananda Chakrabarty.

†If the application that is currently under consideration by the commissioner is refused,
the company that is involved intends to appeal to the Supreme Court.

The Patents Act 1977 (U.K.) came into force in England on June 1, 1978.[118] It dispensed with the "manner of new manufacture" test and lends support to the commissioner's decision to patent the new variety of *Pseudomonas*. The act replaces the "new manner of manufacture" test with a section that provides that a patent may be granted for an invention if the invention is new, involves an inventive step, is capable of industrial application, and is not excluded by certain specified provisions of the act.[119] An invention is not defined, although a list of exclusionary categories is set out.*

Subsection 3(b) is of particular significance, although in one sense it merely constitutes a codification of existing British Patent Office practice. It stipulates that "a particular patent shall not be granted . . . for any variety of animal or plant or any essentially biological process for the production of animals or plants *not being a microbiological process or the product of such a process*" [emphasis added].

The Chakrabarty application has been approved in the United States as well as the United Kingdom, although an appeal to the Supreme Court is pending.[120] At first instance the U.S. Patent and Trademark Office Board of Appeals affirmed the Patent Office rejection of the relevant claims. The board was of the opinion that the issue for decision was whether section 101 of the Patents Act 1952 included "living organisms."[121] Section 101 provides that "whoever invents or discovers any new and useful process, machine, manufacture or composition of matter, or any new and useful improvement thereof, may obtain a patent therefor, subject to the conditions and requirements of this title." In deciding whether those words and, more particularly the word "manufacture" could apply to organisms, the board, after reviewing a number of authorities,[122] stated that there was no case dealing directly with the point in issue in the *Chakrabarty* appeal. It then proceeded to place considerable emphasis on the Plant Patent Act 1930 (U.S.) and the cases that surround it.[123] The board concluded that the passage of that act served as an indication that Congress intended to make separate provision for the patentability of living things and accordingly, it held that section 101 does not include living organisms.

The second relevant application to come before the board was filed on June 10, 1974, by Bergy, Coats, and Malik of the Upjohn Company.[124] The specification related to a biologically pure culture of the microorganism *Streptomyces vellosus*, which was said to be capable of producing significant quantities of the antibiotic lincomycin. These claims were allowed by the examiner in accordance with the established practice of patenting processes that involve the use of organisms.[125]

---

*For the purposes of this book, one of the more interesting of these categories relates to a "mathematical method."

The board rejected the appeal on the basis of the same reasoning that it had applied in the *Chakrabarty* case, namely, that the absence of an express mention of living things in section 101 and the passage of the Plant Patent Act 1930 can be interpreted as a clear indication that Congress did not intend section 101 to apply to organisms. The board then proceeded to consider the consequences of a decision in favor of patentability. In this context the majority stated that "if we were to adopt a liberal interpretation of 35 U.S.C. 101 new types of insects, such as honey bees, or new varieties of animals produced by selective breeding and crossbreeding would be patentable."[126] The dissenting board member placed his emphasis on the words used in section 101. He was of the view that there was nothing in the section that precluded organisms from being patented.[127] On October 6, 1977, the *Bergy* case reached the Court of Customs and Patent Appeals.[128] The court rejected the decision of the board, although it was more circumscribed in its statement of the issue than the board had been. It stated that "we are not deciding whether living things in general or, at the most, whether any living things other than microorganisms, are within section 101. These questions must be decided on a case-by-case basis. . . ."

In deciding that the organism was patentable under section 101, the majority quite rightly emphasized that it would be illogical to insist that the presence of life in a manufacture or a composition of matter removes it from the category of patentable subject matter while recognizing that the use of an organism and its life functions does not affect patentability.[129] The court then referred to the fact that there is nothing in section 101 that excluded live manufactures and held that it was in the public interest to include microorganisms within the terms "manufacture" and "composition of matter" in section 101. Unfortunately, the court did not take that point any further, although it did suggest that "as for the board's fears, that our holding will of necessity, or 'logically,' make all new, useful and unobvious species of plants, animals and insects created by man patentable, we think the fear is far-fetched."[130]

In spite of the fact that it was the first case to reach the Board, the *Chakrabarty* case was not decided by the Court of Customs and Patent Appeals until March 1978. The respondents attempted to distinguish the *Bergy* decision by arguing that in that case the court had limited its judgment to claims involving only biologically pure cultures.* That argument was not accepted on the basis that there was no significant difference between the two claims. *Bergy* was therefore viewed as a controlling precedent. The court was content to simply refer back to the result in the *Bergy* case and to

---

*The same judges were on the bench in both cases.

use that as the sole ground for reversing the decision of the board. In this case the dissenting judges did more than repeat the reasoning of the board. The most noteworthy dissent was delivered by Judge Baldwin, who criticized the classic twofold distinction between products of nature and statutory subject matter. He then referred to a third category that was intermediate between the other two. Into this class fell "things sufficiently modified so as to be statutory 'manufactures.' " Judge Baldwin was of the opinion that the organisms that were specified in the *Chakrabarty* application were within this category and he cited the decision in *American Fruit Growers Inc.* v. *Brogdex Co.* as an authority for that proposition.[131] In that case the court had held that in order to become statutory subject matter an article had to possess "a new or distinctive form, quality or property." Clearly, then, the issue in the *American Fruit* case is distinguishable from the one that is raised by the Chakrabarty application. Chakrabarty's microorganism had been acknowledged to be new and unobvious. However, Judge Baldwin applied the *American Fruit* rule in the context of a discussion of "living organisms" and a supposed novelty issue. He stated that the essential nature of the item that the appellant had modified was its "animateness" or life and that the appellant had not changed its essential nature. Accordingly, in his view the organism was unpatentable. This was clearly the "living organism" issue but the judge then confused the point by adding that the appellant had not created a new life. In this way he succeeded in raising the novelty issue that was at the heart of the *American Fruit* decision but that was irrelevant on the facts of the *Chakrabarty* case.

Unfortunately, the minority judgments of the Court of Customs and Patent Appeals in the Chakrabarty and Bergy applications have been supported in an appeal from the *Bergy* decision. On June 27, 1978, the Supreme Court accepted a writ of certiorari from the Count of Customs and Patent Appeals concerning *In re Bergy*. It vacated the judgment of the court and remanded it back to that court for consideration. The Supreme Court was of the opinion that since the number of living things is vast the decision opens an enormous range of subject matter to patentability. Clearly, the Supreme Court did not accept that the floodgate argument was as "far-fetched" as the Court of Customs and Patent Appeals had suggested. It emphasized its decision in *Parker* v. *Flook,*[132] which concerned the question of whether a mathematical formula is patentable under section 101 of the Patent Act.[133] That case appears to be significant in this context only in relation to the following passage: "It is our duty to construe the patent statutes as they are now read, in light of our prior precedents, and we must proceed cautiously when we are asked to extend patent rights into areas wholly unforeseen by Congress."

Although the decision of the Supreme Court in the *Bergy* case has not yet been reported, it would appear that the Court adopted the cautious approach that had been recommended in *Parker* v. *Flook*. If the court was truly concerned with construing statutes as they now read, it would have been well-advised to concentrate on the words of section 101 and the test for patentability that hinges upon the meaning of manufactures and compositions of matter. Had it adopted that approach its decision might well have been different.[134]

Is the timidity of the U.S. Supreme Court likely to have an influence on the determination of the Chakrabarty and University of California claims in New Zealand? In this country the all-important definition is that of an invention. The *Swift* case stands as authority for the patentability of biological processes,[135] but the case law does not extend to a discussion of the patentability of organisms.[136] Even if the commissioner is prepared to concede that novel genetic techniques, and, more particularly, novel organisms, constitute inventions, he can still argue that he has a discretion to refuse a patent under section 17 of the Patents Act 1953.

Section 17 sets out three major categories of application for which the commissioner has a discretion to refuse patents for inventions. The first category is not relevant. It relates to "any application for a patent if it appears to the Commissioner that it is frivolous on the ground that it claims as an invention anything obviously contrary to well-established natural laws." The second class of application is arguably more relevant. It concerns inventions the use of which "would be contrary to law or morality." Some writers have already suggested that genetic techniques are immoral in the sense that they constitute a god-like interference with nature.[137] However, definitions of morality are necessarily subjective and it would be most unfortunate if these objections were to find legal expression in section 17 (1) (b) of the Patents Act. Even if morality can be said to be represented by the so-called common conscience, it is not easy to argue that novel genetic techniques are immoral in any conventional sense. On the other hand the products and processes of genetic engineering have the potential to cause serious physical harm to man and his environment,* and this is the major justification for controlling genetic engineering by statute. In this context the legal limb of the "law or morality" test becomes relevant. If a genetic engineering act is passed and the technique described in the patent application does not comply with the statute, the commissioner could refuse a patent under Section 17 (1) (b) of the Patents Act.† However, it would clearly be quite impractical to place the primary onus—that of determining whether statutory requirements had been fulfilled—on the Patent Office. That function should be discharged by the proposed National Advisory

---

*See Chapter 2, section entitled "Should Genetic Engineering Be Controlled?"
†If the act were breached after the patent was granted, then the patent could be revoked.

Committee, which would, in practice, have considered the licensing of the relevant genetic engineering projects prior to the filing of related patent applications concerning completed research. Accordingly, one of the duties of the National Advisory Committee would be to advise the Patent Office of its licensing decisions. The problem of confidentiality would hopefully be obviated by the inclusion of an unjustified disclosure provision in the new legislation.

Even in the presence of such a section the situation would be aggravated by the committee membership of experts in the relevant fields. Thus it might be necessary to include additional safeguards in genetic engineering legislation. For example, it would be possible to model a provision on section 6 of the Patents Act. That section precludes officers and employees of the Patent Office from preparing specifications for inventions and from acquiring an interest in any patent within a period of one year after their association with the office.* Sir Gordon Wolstenholme, chairman of the British Genetic Manipulation Advisory Group, has devised what appears to be a practical solution to some of these problems. He has asked GMAG members to declare their business connections, including consultancies and shareholdings. In appropriate cases, certain members can then be excluded from all deliberations that relate to "sensitive" proposals.[138] Section 17 (1) (c) of the Patents Act is also of interest. It gives the commissioner a discretion to refuse a patent for an invention if it appears that an applicant "claims as an invention a substance capable of being used as food or medicine which is a mixture of known ingredients possessing only the aggregate of the known properties of the ingredients or that it claims as an invention a process producing such a substance by mere admixture. . . ."

It is possible that this paragraph could be applied to genetically engineered organisms.[139] They can be said to constitute a mixture of known ingredients and indeed their usefulness lies in the fact that they consist partly of genetic material that is gathered from existing organisms and recombined in the novel organism.† This interpretation would depend to a large extent on the word "ingredient," which is not defined in the act. If past performance can be used as a reliable indicator for the future, it is conceivable that the commissioner will rely on this provision in order to exclude patent applications that relate to novel organisms. Although statutes are always speaking, it is clear that novel genetic techniques were not within contemplation when

---

*It should also be noted that, by virtue of section 60 (1) of the Patents Act, disclosure to committee members would not constitute anticipation or prior publication of an invention for the purposes of the Patents Act.

†This statement does not relate to all novel genetic techniques. For example, the "shotgun" experiment does not, in the truest sense, involve preselection.

section 17 was drafted. The section appears to have been designed to lend statutory support to the principles of novelty and obviousness and should not be used for the purpose of denying patent protection to products and procedures that, despite their component parts, are neither obvious nor lacking in novelty.

In New Zealand, as in England, genetic engineering techniques and genetically engineered organisms would seem, prima facie, to be patentable in the sense that they can be said to constitute inventions in terms of the definition in section 2 of the Patents Act. However, that statement must be qualified not only in terms of section 17 but also in the light of the extremely restrictive interpretations that have been placed on the definition of an invention in New Zealand—restrictions that have no apparent foundation in governing legislation.

The technology of genetic engineering is unlikely to achieve its full promise in the absence of financial and intellectual input from the private sector.[140] That support will not be forthcoming if the products and processes of the new technology are declared to be unpatentable. Patentability also precludes secrecy in a sphere in which openness is vital. On the other hand, it can be argued that all the benefits that flow from genetic engineering techniques should be vested in the community that may ultimately be forced to bear the responsibility for environmental damage caused by the technology.* It can also be claimed that the resource involved is so important that its application in the public interest must be ensured by public sector control.

Such considerations are already recognized in certain sections of the Patents Act. For example, section 55(1) provides that any government department and any person authorized in writing by a government department may make, use, and exercise any patented invention "for the services of the Crown." Conversely, section 55(7) acknowledges that the public interest might best be served by allowing the private sector to work patents that have been developed in the course of governmental research. However, any such use is deemed to be a use for the services of the crown. Section 51 is also relevant. It empowers the commissioner to grant compulsory licenses to third parties in relation to private sector patents over substances that are capable of being used as food or medicine or in the production of food or medicine or the processes for producing such substances.[141]

In conclusion, it can be said that novel genetic organisms and techniques that have already been patented in England are also patentable in New Zealand. The decision of the U.S. Supreme Court in the *Bergy* case was uncharacteristically timid. It will discourage industrial investment in this research and will deprive the community of a much-needed source of infor-

---

*See the next section for a discussion of liability and compensation.

mation about genetic engineering research. In determining the question of patentability the courts have not, in the past, waited for additional legislative guidance. In the U.S. context, however, such guidance may now be necessary in the form of a statutory reinstatement of the judgments of the Court of Customs and Patent Appeals in the *Bergy* and *Chakrabarty* cases.

Civil Liability and Compensation

It is clear that the public cannot effectively participate in the implementation of proposed genetic engineering legislation unless it has access to information — information that will enable it to assess the technology of genetic engineering, to make informed decisions regarding its control, and to institute court proceedings if that course of action is necessary. There are, however, other important community needs that should not be ignored in the context of a discussion of genetic engineering. For example, those who suffer loss have a need for compensation and the community has an interest in making certain that they receive it. But can and should proposed genetic engineering legislation ensure that, even in the event of a mass accident, members of the public who have sustained loss as a result of the manufacture, marketing, or use of genetically engineered organisms are compensated?

This question must, in the first instance, be considered in terms of existing remedies.* The Accident Compensation Act 1972 does not, in general, appear to assist those who suffer losses that are caused by novel genetic organisms or processes. Unfortunately, such losses are unlikely to be viewed as having arisen either directly or indirectly from a personal injury by accident within the meaning of the act.[142] Accordingly, in order to recover their losses injured parties would have to bring a common law action on the basis of negligence, nuisance, or the doctrine in *Rylands* v. *Fletcher.* The difficulties that are inherent in establishing those causes of action and the often unsatisfactory nature of the remedies that are granted by the courts have already been examined in a previous chapter but, at this point, it should also be noted that in practical terms the amount of compensation that is recoverable is determined by the financial status of defendants who are not likely to be insured against legal liability. Thus in the event of a mass accident many plaintiffs will not be able to recover their losses at common law. However, even if damage that resulted from a genetic engineering accident could be said to constitute a personal injury by accident, the Accident Compensation Act would not provide a total solution in the sense that it does not extend to property damage. In addition, the fund itself is relatively small. A reserve fund of $138 million must meet the cost of personal

---

*See the earlier section in this chapter entitled "Existing Law."

injury claims that have already been filed but have yet to be settled.[143] Against that background the commission has expressed the view that the fund could not cope with claims arising out of a major catastrophe.[144] Even if the act could be extended to cover sickness and the fund enlarged accordingly, that argument would still stand. Agricultural damage would also remain as a problem that would not be likely to be solved by further extensions of the scope of the act.

The establishment of a special contribution-based national compensation fund for the victims of genetic engineering accidents has only limited appeal. The risks that are involved in the technology do not appear to be sufficiently great to justify significant compulsory contribution by a group large enough to make a financial impact. However, the appropriation of public monies for the establishment and maintenance of such a fund may be more appropriate.[145] The Earthquake and War Damage Fund does not provide a suitable model. It is maintained largely by insurers' contributions, which are recovered in the form of inflated fire insurance premiums.[146] That concept may well prove to be unacceptable to a community that will view widespread storm, flood, and earthquake damage[147] as more imminent than widespread loss caused by novel genetic organisms and techniques.*

On the other hand, members of the community will wish to be compensated if an accident occurs. How could they recover their losses in the absence of a contribution-based national fund? It can be argued that those whose actions cause the losses in question are the most appropriate loss bearers. At first sight that suggestion does not appear to represent an improvement on the uncertainties of the existing common law position. However, what is envisaged is statutory strict liability coupled with the imposition of a duty to insure against claims for damage. The concept of imposing strict liability for activities that could result in widespread and serious damage was recognized in the case of *Rylands* v. *Fletcher*[148] and has more recently been accepted in relation to liability for nuclear accidents.[149]

Perhaps strict liability is too harsh a burden to impose. Spigelman has warned against inflexible statutory sanctions that cannot be molded in terms of fault and culpability.[150] Section 40 of the Animals Act 1967 represents an alternative. It deals with the liability of owners of animals that have communicated disease to stock that belong to a third party. The section imposes something less than strict liability in the sense that only those owners who are subject to a penalty under the act can be held to be strictly liable. Spigelman's views are, however, expressed in the course of a discussion of sanctions per se. In this subchapter the concern is with civil liability

---

*The recent events in Birmingham with regard to an escaped smallpox virus may increase public awareness of the risks that are involved.

and compensation and the proposed statutory strict liability would have compensatory rather than punitive or deterrent aims, although a resulting improvement in laboratory and handling procedures would be a welcome bonus.[151] For those reasons and in order to avoid any element of chance or uncertainty in the law it would be unwise to adopt the penalty-related solution in section 40 of the Animals Act[152] or to limit strict liability to injuries caused by "hazardous biological research."[153]

The sponsoring institute rather than the project supervisor would be the legal person to whom liability would attach under the proposed provision, although those who administer the act could check compliance with the insurance requirement by referring to genetic engineering licenses. The license would be issued in the name of the project supervisor in order to ensure that different projects that are conducted under the auspices of a single organization can be effectively monitored but they would also contain the name of the sponsoring institute.

The greatest danger that is involved in forcing organizations to insure against legal liability is that the cost of premiums would be so prohibitive that the technology would be abandoned in favor of more profitable but less beneficial alternatives. If minimum insurance cover of $5 million were prescribed and if this were granted on the basis that there is not a high likelihood that the activities involved would result in losses of that magnitude, then the premiums should not be excessively high. However, the fact that strict liability is imposed would affect the premiums, although probably not to such an extent that they would be likely to be unbearably large as far as the organizations involved in this type of work are concerned. *

Despite the fact that a statutory minimum insurance cover would be prescribed, amounts *in excess* of the minimum figure should still be recoverable, but only at common law.† To take an example, the possibility of a successful negligence action in relation to the full amount of the damage caused would be unlikely to deter investment in genetic engineering research, whereas it can be argued that the imposition of strict liability up to an unlimited amount could have that effect. If this assumption is not correct and those who are involved in the technology of genetic engineering consider that the risk of accident is prohibitive, then the abandonment of the new technology before serious damage occurs would be a substantial step in the public interest.

---

*It is interesting to note that it is not uncommon for New Zealand pharmaceutical companies to take out insurance coverage that is in excess of $5 million. The statements in the preceding paragraph are based on discussions with the N.Z. State Insurance Office.

†Special governmental intervention will be required if there is a mass accident and claims that exceed insurance cannot be recovered at common law either because there is no tort or because the tortfeasor lacks the necessary substance.

Offenses and Penalties

The statutory duty to provide compensation for conduct that causes injury is not the only liability that researchers and organizations would incur under the proposed act. The main difference between control by means of nonmandatory guidelines and informal controls as opposed to statute lies in the formality of the legal sanction. Offenses are created in the knowledge that not only a penalty but also a stigma attaches to conviction. The nature and size of that penalty will vary according to whether the aim is simply punitive or retributive or whether a deterrent goal is present.

Offense and penalty provisions in genetic engineering legislation, and indeed in all statutes, should reflect society's disapproval in order to discourage the prohibited conduct rather than to extract the salutary pound of flesh, although it has been suggested that even the latter function serves a limited social purpose.[154] Certain types of conduct are prohibited because they are perceived to be dangerous or in some way undesirable. The proposed genetic engineering legislation is designed to control a new technology in order to prevent environmental damage. It would achieve that end by providing for the establishment of supervisory committees, the issuing of licenses, and the inspection of premises.* Accordingly, the following would constitute offenses: the misrepresentation of any material fact for the purpose of obtaining a license or for any other purpose in relation to the act;[155] the breach of any conditions of a license; a failure to comply with requests for information; and a breach of any provision of the act or any regulation promulgated under it.† Offenses that relate to the unjustified disclosure of information that is obtained in the course of administering the act have already been considered,[156] and it can be argued that such offenses should appear in the context of sections that deal with confidentiality and disclosure rather than in a general offense provision.

Section 13 of the Summary Proceedings Act 1957 obviates the need to draft a special section for the purpose of ensuring that members of the public have a right to institute proceedings in relation to an offense. The section, which has not been overlooked by concerned interest groups in the past,[157] reads as follows: "Except where it is expressly otherwise provided by an Act any person may lay an information for an offence." This is an important section in the environmental context as it enables vigilant individuals or societies to take action in the event that those who administer an act refuse to do so.[158]

The substance of the offenses and the width of the class of potential prosecutor has been established, but two important questions remain to be

---

*See the earlier section in this chapter entitled "Committee Structures and Functions."

†As, for example, when genetic engineering research is conducted without a license.

considered. First, will conviction be automatic if noncompliance with a section of the act is proved beyond reasonable doubt? Mahon, J. has expressed the view that the "traditional approach to the proof of statutory offenses is obviously too deeply embedded in the legal system to be displaced." However, in the context of new legislation the necessity for adherence to accepted but unsatisfactory past practices is not a cogent reason for a refusal to depart from the old standards and as Mahon, J. points out,

> in a prosecution for conduct involving environmental damage it seems inappropriate, having regard to the public rights placed in jeopardy by the alleged conduct, to apply any standard of proof other than the balance of probabilities. The duties sought to be enforced by such a prosecution will be owed to a considerable section of the community, in some cases to the entire community and it certainly seems sufficient for the purposes of imposing a penalty for breach of that duty, to show that there is a preponderance of evidence in favour of liability.[159]

In terms of the nature of the liability, as opposed to the standard of proof that is required, it is interesting to note that the imposition of strict liability has been proposed in the United States. Thus, if a section of the act is breached by the accused he cannot escape conviction by arguing that the breach occurred unintentionally or without his knowledge.* On the other hand an accused who can be shown to have acted knowingly or willfully is subject to a greatly increased penalty. Such an approach is relatively stringent, although section 56 of the Noxious Plants Act 1978 also imposes strict liability for violations of the provisions of the act.

In the civil context liability without fault can constitute a relatively equitable means of loss distribution in the sense that those who engage in high-risk activities can be expected to cover the cost of them and are often in the best position to do so. However, the primary goal in this particular setting is not to compensate victims but to deter offenders. Against that background it can be argued that the imposition of strict criminal liability will not serve the desired deterrent purpose, indeed, it is difficult to envisage that it would serve any useful purpose. Perhaps a "strict liability" designation would demonstrate the strength of society's views on the dangers of genetic engineering, but such a display would be largely irrelevant because it could not discourage truly accidental behavior that could not have been avoided even by the adoption of safety precautions.

A "half-way house" solution of the kind that is embodied in section 31 of the Food and Drug Act 1969 is arguably less stringent without being less effective. The section provides that in any prosecution for selling food,

---

*Appeals could be handled by existing appellate courts in the usual fashion.

drugs, or medical devices contrary to the provisions of the act or regulations it is not necessary for the prosecution to prove that the accused intended to commit an offense. On the other hand the accused will have a good defense if he can prove that he did not intend to commit an offense against the act or regulations and that he took all reasonable steps to ensure that his actions would not constitute an offense.[160]

The remaining issue concerns the penalties that would be imposed under the act. Is imprisonment appropriate? If a fine is preferable, how large should it be and on whom should it be imposed? It is in this context that Spigelman's warning is particularly relevant.[161] The legislation should confer a wide but not unlimited discretion to tailor penalties to fit the seriousness of the crimes that are involved. The possibility of the revocation, suspension, or limitation of licenses would seem likely to act as an effective deterrent.* Imprisonment smacks of mere retribution, although it may have a deterrent effect. Fines would seem to constitute a much more appropriate penalty. They could also be used as a preventive measure. Thus a maximum penalty of $10,000 could be imposed for violation of any provision of the act. Each day for which the breach continues could be deemed to give rise to a separate violation for the purposes of the section.† Because prevention is the most important goal, the separate daily penalties for continuance should not be decreased below the level of the initial fine, although such a reduction is not uncommon, even in situations where continued offending is clearly damaging.[162] Presumably the lesser fine represents an acknowledgement of the fact that it is often very difficult to remedy a breach quickly and that, in the interim, companies should not be forced out of business. In this context an informed public should be particularly watchful in ensuring that the weight of concentrated corporate power does not upset an imbalance in favor of the public interest. It can also be suggested that a maximum $10,000 fine should be replaced, in the case of a corporation, by a maximum of $100,000.‡

The fines could be imposed on any person, but because of the proposed statutory definition of "a person," all researchers, technicians, groups, corporations, and companies involved in the technology of genetic engineering

---

*If the conditions of a license were breached, the minister, on the advice of the National Advisory Committee, could exercise these powers. However, the courts should also be able to employ these sanctions to penalize breaches that do not relate to licenses. In the former case, the accused should be given notice of an intention to revoke, suspend, or limit the license.

†These penalties could be imposed after conviction on indictment. Compare the approach proposed in the United States where large "civil penalties" would be imposed by administrative agencies.

‡Such a large fine is not unprecedented. For instance, section 81G of the Commerce Act 1975 provides for a maximum of $100,000.

would then be subject to these penalties. For that reason all conduct authorized by the licensee or project supervisor, as defined in the act, should be attributed to him, although in his defense he would be able to rely on the fact that he exercised the degree of reasonable care that was required by the situation — the same defense that would be available in respect to his own actions.[163] The project supervisor's employer and all company directors should be subject to an automatic penalty if the project supervisor was held to have committed an offense. In that sense the statute would impose vicarious criminal liability up to a $10,000 maximum in the case of individual directors. Hopefully, such provisions would stimulate potentially lackluster employer safety committees to perform a vigilant and valuable role in enforcing the proposed legislation.

NOTES

1. It did not prevent researchers at Harvard and the University of California from flouting the NIH guidelines. On the other hand, John Carbon, a research biologist at the University of Santa Barbara, recently destroyed the products of two years' work because he was informed by his colleagues that the organisms he had produced as a result of "shotgun" experiments were proscribed by the NIH guidelines.See *Nature* 226 (1977): 210.

2. P. Spigelman, "Sanctions, Remedies and Law Reform," unpublished Australian Law Reform Commission Discussion Paper.

3. These arguments have also been raised in relation to public participation in decision making (see Chapter 2). It is notable that scientists in New Zealand are not controlled by their own professional society. They have neither a code of ethics nor a disciplinary committee that can review the actions of individual researchers. A "Code of Practice for Scientists" was published in 1972 but it was concerned more with the rights than with the responsibilities of scientists. See NZ *Science Review* 30 (1972): 33. In fact paragraph 2.1 of the code suggests that "scientists should accept the principle that the standards of science and the status of scientists are determined by the scientists." None of the provisions of the code relate to enforcement or to a regulatory body. When introducing the code the council of the New Zealand Association of Scientists quite aptly described its effort as a "service for scientists," although even that is questionable.

4. For example, see R. Curtiss, "An Open Letter to Donald Frederickson, Director of the National Institute of Health," April 12, 1977.

5. It is interesting to note that the Royal Society of New Zealand has established a new two-member standing committee that is to advise the society's 43 special interest groups of the introduction of any relevant bills. Hoare and Barber are the first appointees. See "Microscope on Politics," *The Dominion*, November 2, 1978, p. 2.

6. See the Radiation Protection Act 1965 and the Radiation Protection Regulations 1973.

7. A nuisance action does not have to be private. The attorney-general can bring a public nuisance action if he is of the opinion that the public interest has been threatened. If the attorney-general refuses to intervene in government-sponsored research, then a plaintiff who can show that he has suffered particular damage, over and above that which is sustained by the public at large, can bring a public action. Recent indications of willingness to extend *locus standi* are encouraging. See *Attorney-General* v. *Independent Broadcasting Authority* [1973] QB 629; *Gouriet* v. *Union of Post Office Workers* [1977] 2 WLR 310 (C.A.); *Harder* v. *NZ Tramways and Public Passenger Transport Authorities Employees Industrial Union of Workers* [1977] 2 NZLR 162. However, compare *Gouriet* v. *Union of Post Office Workers* [1977] 3 All ER 70 (H.L.).

8. Plants Act 1970, section 4.

9. Ibid, section 12. Compare the Public Safety Conservation Act of 1932, section 2.

10. Ibid, section 11.

11. Ibid, section 6.

12. Ibid, section 7.

13. Ibid, section 19.

14. Forests Act 1949, section 70.

15. Ibid, section 70(3).

16. Animals Act 1967, section 2.

17. Health Act 1956, sections 10, 9A, 20, and 28.

18. In *Pyrah* v. *Doncaster* (1949) 41 BWCC 225. Bucknill, L.J., in the context of a discussion of accidents and diseases, stated that he could not distinguish between incursions into the body of living organisms, and mineral bodies such as fragments of silica dust.

19. For example, see sections 10, 11, 13, 16, and 17 of the Clean Air Act 1972.

20. Ibid., section 6.

21. Ibid., section 48. Members of the public can lay informations under section 13 of the Summary Proceedings Act 1957 if the terms of a license are breached.

22. Ibid., section 52.

23. The defects in the genetic engineering regulations that were promulgated under the Health and Safety Act 1974 (U.K.) have already been discussed. Even if the British approach were thought to be desirable, a similar course of action could not be adopted in New Zealand because, in that country, provisions relating to safety at work are scattered through several different statutes and do not extend, in the same way that their British counterparts do, to public safety. See *Labour Department* v. *Merrit Beazley Homes Ltd* [1976] 1 NZLR 505, 508 and also the Factories Act 1946, the Shops and Offices Act 1955, and the Accident Compensation Act 1972.

24. *Donoghue* v. *Stevenson* [1932] AC 562, 580.

25. John G. Fleming, *The Law of Torts* (Sydney: Law Book, 1971), p. 136.

26. Op. cit., 562.

27. *Hedley Byrne* v. *Heller* [1964] AC 465, 534.

28. [1966] 1 QB 569.

29. *Weller*, op. cit., p. 587. The limitation to cattle in the neighborhood goes to the question of remoteness of cause, although the court did not specifically proceed beyond the issue of whether a duty existed.

30. Ibid., p. 577.

31. For example, see *Delisle* v. *Shawinigan Power Co.* (1969) 4 DLR 3d, 458.

32. The plaintiffs in *Environmental Defence Society* v. *Agricultural Chemicals Board* [1973] 2 NZLR 758, 762 were faced with similar difficulties. In that case Haslam noted that "it was conceded by counsel in argument that the tragic affliction suffered by the child of the member of the Society had not been proved to have been caused by contamination from this chemical." The Health Department has also denied that spraying activities caused a series of miscarriages in the Waikato and, more recently, in South Taranaki.

33. Compare the "direct consequence" test that is discussed in Fleming, op. cit., p. 179.

34. (1962) 80 WN (NSW) 852.

35. [1962] 2 QB 405.

36. (1866) LR 1 Ex, 265, 279-80.

37. At the New Zealand Law Society's Triennial Conference in March 1978, McMullin, J. expressed the view that the rule in *Rylands* v. *Fletcher* could be applied to the escape of the products of genetic engineering research "provided always that the ingenuity of the courts can be counted upon." See "Conference Courier," August 28, 1978, p. 1.

38. Compare *Longmeid* v. *Holliday* (1851) 6 Ex. 761 and *Faulkner* v. *Wischer and Co.* [1918] VLR 513. In those cases the courts discussed liability for negligence in terms of things that were "inherently dangerous." However, the decisions can be challenged on the basis that, in terms of the tort of negligence, liability attaches not to any activity involving inherently dangerous substances, organisms, or products but to the negligent acts surrounding their manufacture, handling, or release.

39. The plaintiff in the *Weller* case argued in the alternative on the basis of the rule in *Rylands* v. *Fletcher,* although that claim was dismissed because the defendants had no proprietory interest in the cattle market into which the disease escaped. See also *Cattle* v. *The Stockton Waterworks Co.* (1875) L.R. 10 QB 453.

40. [1947] AC 156.

41. See Fleming, op. cit., p. 286, and F. H. Newark in 'The Boundaries of Nuisance" (1949) 65 LQR 480, 488.

42. *The Wagonmound No. 2* [1966] 2 All ER 709 and *Goldman* v. *Hargrave* [1967] 1 AC 645. The plaintiff who pursues an action in nuisance has an advantage in terms of injunctive relief.

43. Accident Compensation Act 1972, section 5.

44. Ibid., section 2 (b) (ii). Although it could be argued that novel recombinant organisms cannot produce disease or infection in any conventional sense of those words, it is clear that under the Workmen's Compensation Acts an infection could constitute an injury by accident for the purposes of the legislation. For example, in the case of *Pyrah* v. *Doncaster,* op. cit., it was held that infection by tuberculosis bacilli was an accident arising out of employment within the meaning of the Workmen's Compensation Act 1925 (U.K.). After expressing the view that the distinction between accident and disease was well-founded, Bucknill, L.J. (dissenting) went on to say that he could not distinguish between "injurious incursions into the body of living organisms, chemical bodies such as particles of gas and mineral bodies such as minute fragments of silica dust." Cohen, L.J. noted that although there was no reported case in which a workman who had been infected by inhalation had recovered, there was also no authority that suggested that workmen could not recover in such a case. He then reviewed the relevant decisions in this area (see *Pyrah,* op. cit., p. 267). The foremost of these were *Brintons Ltd* v. *Turvey* [1905] AC 230 and *Innes (or Grant)* v. *Kynoch* [1919] AC 765. In the *Brintons* case an anthrax bacillus had entered the plaintiff's body through his eye and this infection was held to be an injury by accident. In the *Innes* case the court found in favor of the widow of a workman who had been infected by noxious bacilli that entered his body through an abrasion on his leg. In that case Lord Birkenhead, L.C. based his decision on *Brintons* and noted, at page 770, that the number of diseases that can be traced to infection by bacillus has increased with the result that the decision in the *Brintons* case was likely to be of wider application than might first have been imagined. However, Lord Birkenhead then proceeded to emphasize that "no apprehension founded upon these scientific observations can affect our duty to follow the decision when once we are agreed upon its scope." Finally, Denning, L.J. who, like Cohen, L.J., accepted the reasoning in *Brintons* and *Innes,* held that "many diseases cannot in their nature be due to accident; but some may be. These include the diseases produced or precipitated by trauma, that is, injury or shock; by exposure or exertion; by infection or contagion; or by contact with or invasion by foreign elements" (see *Pyrah,* op. cit., p. 270).

45. Decision No. 50, Reference No. 65/76, July 6, 1977; (1977) 1 NZAR 295.

46. Ibid., p. 297.

47. Decision No. 2, Reference No. 21/2/7, May 29, 1975; (1976) 1 NZAR 43.

48. Ibid., p. 43.

49. For instance, in *Russell Transport Ltd.* v. *Ontario Malleable Iron* [1952] 4 DLR 721, 723, the plaintiff successfully relied on expert evidence in order to prove that pollutants emitted into the air from the defendant's foundry caused paint damage to cars in the plaintiff's adjacent yard. It was noted that when a chemist inspected the cars he found particles that were determined through microscopic examination to be "red iron rust, black iron scale, white cast iron, grey cast iron or malleable pearl cast iron particles some of which were spherical in form and manganese sulphide crystal, . . . particles incident to foundry operations," p. 723.

50. Accident Compensation Act 1972, section 67(2) (d).

51. It cannot be disputed that infection or disease that arises out of the physical consequences of a personal injury by accident or as a result of medical or surgical misadventure is a personal injury by accident under section 2. See the opening words in section 2(1) (b), which deals with exclusions. The relevant words are "Except as provided in the last preceding paragraph. . . ." See also Review Decision No. 77/R.2199, although it was held that there was no medical or surgical misadventure in that case.

52. As, for example, in the type of situation in *Smith* v. *Leech Brain* op. cit., p. 151.

53. "Report of the Working Party on Novel Genetic Techniques," April 1978, p. 7.

54. For example, see Curtiss, "An Open Letter to Donald Frederickson . . . ," op. cit.

55. "Report of the Working Party on Novel Genetic Techniques," op. cit. That phrase was also used by the MRC Advisory Committee.

56. For example see *8 o'clock*, May 21, 1977, p. 10.

57. See the Report of the MRC Advisory Committee, "Recommendations of the MRC Advisory Committee on Genetic Manipulation," 1977 (unpublished), p. 2. In this context the British choice of genetic manipulation is perhaps a little surprising.

58. See "Frankenstein Monster Could Create New Form of Death," *Evening Post*, November 26, 1976, p. 27.

59. See the "Report of the Working Party on Novel Genetic Techniques," 1978, and the Parliamentary Paper G. 21A, p. 7. The control of microbiological hazard would involve the regulation of all activities that involve microorganisms.

60. See the "First Report of the Genetic Manipulation Advisory Group," HMSO Cmnd. 7215, p. 5.

61. "Novel Genetic Techniques in New Zealand," op. cit., p. 28.

62. *Occupational Health, Laboratory Safety for General and Microbiological Laboratories*, N.Z. Department of Health, January 1973.

63. Unfortunately, the Cabinet Committee that drafted a code of procedure for the Commission for the Environment in 1972 did not include guidelines for the audit of impact reports on proposed environmental legislation and regulations, and the commission has taken the view that its role does not extend to the assessment of such reports. On the other hand environmental impact reports are required from government-funded bodies if it appears that their proposals have significant implications for the "human, physical or biological environment." Theoretically, this could involve government-funded institutes in the preparation of environmental impact reports on important genetic engineering projects. For further details of the reporting process, see *A Guide to Environmental Law in New Zealand* (Commission for the Environment, 1976), p. 103. In May of this year the New Zealand cabinet approved some minor changes in terms of the role and procedures of the commission. However, the amendments contain no mention of the preparation or auditing of impact reports on legislation. See Ian Baumgart, "Environmental and Enhancement Operations," An Inter-Departmental Circular, September 12, 1978.

64. The Environmental Council is described in *A Guide to Environmental Law in New Zealand*, op. cit., p. 104. The council's views on public access to information are set out in the "Report of the Commission for the Environment For The Year Ended March 31, 1978," Paper C.7, Presented to the New Zealand House of Representatives by Leave, p. 15.

65. The Irvine Committee was of the opinion that only "commercial firms" should be licensed, although reasons for the decision to exclude government-funded institutes from the licensing process were not given. See the "Report of the Working Party on Novel Genetic Techniques," op. cit., p. 5.

66. Compare the experience of the GMAG in the United Kingdom as set out in "First Report of the Genetic Manipulation Advisory Group," HMSO Cmnd. 7215, May 1978, pp. 3.16-3.18. Before making an unfavorable recommendation the Advisory Committee would notify the person concerned and provide him with an opportunity to appear in person to present further information and to be represented by a counsel or agent. As to this point and general matters of procedure, see K. J. Keith, A Code of Procedure for Administrative Tribunals, Legal Research Foundation Occasional Paper No. 8., 1974; and "Administrative Tribunals Constitution, Procedure and Appeals," Sixth Report of the Public and Administrative Law Reform Committee, March 1973. See also Radiation Protection Act, section 2(4).

67. See the "Report of the Working Party on Novel Genetic Techniques," op. cit., p. 5.

68. "Report of the Working Party on Novel Genetic Techniques," op. cit., p. 5.

69. The institute has already formulated a code of laboratory practice. See Laboratory Safety for General and Microbiological Laboratories, January 1973.

70. Op. cit.

71. See the "Report of the Working Party on Novel Genetic Techniques," op. cit., p. 6.

72. "Recommendations of the MRC Advisory Committee on Genetic Manipulation," op. cit., 4.

73. Recombinant DNA Bill S. 1217, April 1, 1977, 95th Cong., 1st sess., clause 7.

74. See, for example, "Environment Agency used by State as 'White Wash,' " Evening Post, July 24, 1978.

75. In response to a question from Richard Prebble (M.P.) concerning the likelihood of the introduction of a bill to give statutory independence to the commission, Bill Young replied that "the government has considered the issue and has decided not to proceed with legislation at present. I am happy to leave the question in abeyance for the time being. The Commission is operating very effectively without a statutory basis and its independence is widely accepted." See N.Z. Parliamentary Debates 412 (1977) 2214. It is interesting to note that the commission has played an important role in advising government departments on the control of toxic substances. In fact it has also been suggested that there may be a need for the establishment of an independent control agency for toxic substances. See "Report of the Commission for the Environment for the Year Ended 31 March 1978," op. cit., p. 9. See also Baumgart, op. cit.

76. Scientific and Industrial Research Act 1974, section 5.

77. National Research Advisory Council Act 1963, section 7.

78. In particular, see section 24.

79. For an overview of the bureau's position within the Health Department, see A Guide to Environmental Law in New Zealand, op. cit. p. 109.

80. Animals Act 1967, section 30(4)(a). These powers can be delegated to persons whom the minister authorizes to act on his behalf.

81. This provision is similar to section 11 of the Radiation Protection Act 1965. In addition to the extensive regulation-making powers that are conferred on him under section 31 of that act, the minister of health is empowered to dispose of radioactive material. See also sections 11, 13, and 14 of the Plants Act 1970 and sections 33, 34 and 35 of the Animals Act, which relate specifically to the treatment and destruction of diseased or infected animals. Compare the Forests Act 1949, section 70.

82. Dangerous Goods Act 1974, section 33 (3).

83. Bill Sutton, "Genetic Engineering in New Zealand," lecture delivered to the New Zealand Institute of Chemistry at Hamilton, August 24, 1977, p. 7.

84. See section 4 (1) of the Judicature Amendment Act 1972, although the section does not refer to the interest of the applicant for review. As an alternative to the institution of proceedings for a writ of mandamus, an interested member of the public may, in certain circumstances, be able to enlist the assistance of the Ombudsmen who would find their jurisdiction in section 13(1) of the Ombudsmen Act 1975. The section relates to "a matter of administration" that would appear to include a failure to administer. See case No. 719, *The Ombudsman's Report* 35 (1964), cited in Walter Gellhorn, *Ombudsmen and Others; Citizens' Protectors in Nine Countries.* (Cambridge, Mass.: Harvard University Press, 1966), p. 107. This course of action is open only to an individual if, under section 13 (7) (1) of the act, there are special circumstances that render resort to other remedies "unreasonable."

85. For a very interesting discussion of the burden of proof in environmental cases see David Williams, "Environmental Law — Some Recurring Issues," *Otago Law Review 3* (1973-76): 372, 383 et seq.

86. Compare The Recombinant DNA Standards Bill, S.621, clause 12, op. cit., p. 81. The New Zealand courts have demonstrated that they are willing to assess potential environmental hazards in terms of the public interest. In *Mundy* v. *Cunningham* (1973) NZLR, 557, Haslam, J. expressed some interesting views in the context of common law injunctions and the Town and Country Planning Act 1953. He stated that "the speedy and irreversible methods of modern engineering both in excavation and demolition, leave an instant application to the court as the only method whereby the status quo can be preserved, at least for a brief period to enable a proper investigation of the competing contentions." Note that under sections 36 and 37 of the Town and Country Planning Act 1953 the right to seek injunctions is conferred on local bodies rather than on citizens. Nevertheless it must be remembered that there is a statutory prohibition on granting injunctions against the crown. See sections 17 (1) (a) and (2) of the Crown Proceedings Act 1950.

87. A desirable general reform of the law relating to *locus standi* has already been proposed by the Public and Administrative Law Reform Committee. The committee is of the view that a uniform standing requirement should apply to applicants for review under section 4 of the Judicature Amendment Act 1972. The recommended standard is "a sufficient interest in the subject matter of the application." At present various tests are invoked according to the remedy that is sought. See the majority opinion of the committee in its eleventh annual report, "Standing in Administrative Law," March 1978. Some of the issues involved in this area are also reviewed in "Access to the Courts — Standing: Public Interest Suits," Australian Law Reform Commission Discussion Paper No. 4, 1977.

88. [1973] 2 NZLR, 758.

89. See *Attorney General* v. *Independent Broadcasting Authority*, op. cit.; *Gouriet* v. *Union of Post Office Workers*, op. cit.; and *Harder* v. *N. Z. Tramways and Public Passenger Transport Authorities Employees Industrial Union of Workers*, op. cit. For the reversal, see *Gouriet* v. *Union of Post Office Workers* [1977] 3 All ER, 70 (H.L.).

90. Op. cit., p. 762.

91. S.945, March 8, 1977, 95th Cong., 1st Sess.

92. California Legislature, No. 757.

93. For a very full discussion of the issues involved, see K. J. Keith, "Appeals from Administrative Tribunals," *Victoria University of Wellington Law Review* 5 (1968-70): 123.

94. Ibid., p. 162.

95. *Environmental Defence Society Inc.* v. *Agricultural Chemicals Board*, op. cit.

96. See section 21 (4). The same approach is adopted in the Radiation Protection (Appeals) Regulations, S.R. 1974/319.

97. Note that, in the first instance, there is a right of appeal to the director-general. See section 32.

98. See Clean Air Act 1972, section 35.

99. As to the latter points, see Keith, "Appeals from Administrative Tribunals," op. cit. pp. 164 and 166.

100. See Ombudsmen Act 1975, section 13 (7) and note 84 of this chapter. Grafstein refers to the possibility of appointing a technological ombudsman but he concludes that technology is "too pervasive" to be surveyed by a central agency. See J. S. Grafstein, "Law and Technology, a Technological Bill of Rights" *Canadian Bar Review* 51 (1973): 221 at 244.

101. See John L. Robson, *The British Commonwealth — The Development of its Laws and Constitutions* (London: Stevens, 1967), pp. 189-90 for a more detailed discussion of the relevant preconditions.

102. As Keith points out, the ombudsmen can disclose certain departmental information to complainants. See K. J. Keith, "Constraints on Freedom of Dissemination of Scientific Knowledge," *New Zealand Law Journal* (1976): 512,514. The potential role of a Commonwealth ombudsman in relation to proposed Australian Freedom of Information legislation is described in "Policy Proposals for Freedom of Information Legislation," Report of Interdepartmental Committee, Australian Attorney General's Department, November 1976, paragraph 21.11.

103. For discussion of these provisions, see Keith, ibid.

104. See ibid., p. 516; Dennis K. Clifford, "Freedom of Information: A Legislative Impossibility?" *Victoria University of Wellington Law Review* 9 (1978): 451; William Hodge, "Freedom of Information in New Zealand," *Recent Law* 4 (1978): 284; and "Freedom of Information," a report by the Auckland District Law Society Public Issues Committee, March 9, 1978. The Royal Society of New Zealand has called for "the creation of a New Zealand Freedom of Information Act along with entailed changes in New Zealand laws such as the Official Secrets Act and the State Services Act," See the Royal Society of New Zealand *Newsletter* (Otago Branch), May 1977.

105. At the request of the minister for the Environment, the Environmental Council is currently preparing a report on public access to information. See *The Dominion*, March 26, 1977, and the "Report of the Commission for the Environment for the Year Ended 31 March 1978," op. cit., p. 15.

106. Leslie W. Gandar, "Science Policy," *N.Z. Science Review* 33, no. 6 (1976): 122, 127. Gandar's view can be contrasted with that of the assistant director-general of the DSIR, who has stated that "the subject of genetic engineering, that is, the modification of life forms by unorthodox genetic procedures, is an example of one which is too important not to be reported fully and objectively. It is understandably and quite rightly a subject of considerable public interest and concern today, and the concern is shared by the Department. In DSIR we believe that it is essential that the public should be fully aware of the issues involved in experiments which have been done and are being done, or are contemplated in the general field." See Graham W. Butler, "Administering Science in the Environmental Age," *N.Z. Science Review* 34 (1977): 95,97. Unfortunately this attitude is not reflected in DSIR publications. The only references to genetic engineering research that has been carried out in the department are contained in paragraph-length lists of experiments with novel genetic techniques. There is no attempt to describe the techniques involved or the significance of the research. See *DSIR Research* 189 (1975),; 160 (1976); and 120 (1977). In 1974, the year of the Palmerston North accident, the following statement appeared in the course of a general description of the activities of the Plant Physiology Division: "A long term aim of the Division is to increase the genetic potential of grain and forage crops by overcoming species imcompatability, introducing genetic variation into populations which reproduce without requiring pollination and transferring selected bacterial genes into plant cells." See *DSIR Research*, 174 (1974). The DSIR Annual Reports, which are tabled in the House of Representatives, do not refer to the incident at Palmerston North.

107. Compare paragraph 12.14, "Policy Proposals for Freedom of Information Legislation," op. cit.

108. See *Attorney General* v. *Jonathan Cape* [1976] QB 752.

109. *Distillers Co. (Biochemicals Ltd)* v. *Times Newspapers Ltd.* [1974] 3 WLR 728.

110. Ibid., p. 739.

111. Compare the Health and Safety at Work Act 1974 (U.K.), sections 27 and 28; and section 480 of the Recombinant DNA Bill 1977 (U.S.), H.R. 7418, May 24, 1977, 95th Cong., 1st sess.

112. See J. Gault, "Patent Law and Practice in New Zealand," in *Digest of Commercial Laws of the World — Patents and Trademarks* (National Association of Credit Management) (Dobbs Ferry, N.Y.: Oceana, 1969). This is also the position in Australia. See *National Research Development Corporation* v. *Commissioner of Patents* [1960] ALR 114. That case involved a patent that was granted in relation to the novel agricultural application of known chemicals. See Kenneth Moon, "A Functional View of Copyright, Designs and Patents," *Victoria University of Wellington Law Review* 8 (1976): 303, 308.

113. [1960] NZLR 775. The decision concerned a new method of tenderizing meat by means of the use of enzymes.

114. *Joos* v. *Commissioner of Patents* [1973] RPC 59. See also Moon, op. cit., p. 308.

115. *Commercial Solvents Corporation* v. *Synthetic Products Co. Ltd.* 43 RPC 185.

116. The Patents Act of 1947 (U.K.), section 32. Compare the Patents Act of 1953, section 21 (d) and (e). See also Thomas A. Blanco White, *Patents for Inventions and The Protection of Industrial Designs* (London: Stevens, 1974), Chapter 4.

117. Patent Specification 1,436,573, 4.

118. The legislation appears to be based on "Patent Law Reform," Report of the Department of Trade, April 1975, Cmnd. 6600.

119. Patents Act 1977 (U.K.), section 1 (1). The concept of novelty is retained in section 2 and the test of obviousness is incorporated into section 3 in terms of an inventive step.

120. The appeal is in respect to application 260,563 filed by Ananda Chakrabarty on June 7, 1972. The decision of the board is discussed in *In re Chakrabarty* 197 USPQ 73 (1978). It should be noted that the board accepted that the microorganisms in question were not products of nature and that the appellant had invented strains of bacteria that were new, useful, and unobvious.

121. 35 USC 101. The case relates only to the patentability of the organisms themselves because processes that involve the use of microorganisms are clearly patentable in the United States. See *Dick* v. *Lederle Antitoxin Laboratories* 43 F. 2d, 628 (1930).

122. In particular, see *Funk Bros Seed Co* v. *Kato Co,* 333 US (1948) 127,130. In that case the court held that "he who discovers a hitherto unknown phenomenon of nature has no claim to a monopoly of it which the law recognizes. If there is to be invention from such a discovery, it must come from the application of the law of nature to a new and useful end."

123. See *In re Arzberger* 112F. 2d, 834 (1940), compare section 13 of the Plant Varieties Act of 1973 (N.Z.), which relates to applications for grants of plant selectors' rights in relation to "any new plant varieties." The Plant Varieties Extension Order (No. 21) 1976 extends the act to annual ryegrass, potatoes, field and garden peas, lotus, lucerne, barley, and fodder-type perennial ryegrass. Under section 22 of the act a plant selector has the exclusive right to reproduce or sell the plant varieties to which the grant relates.

124. Application 477, 766. The decision of the board is discussed in *In re Bergy, Coats and Malik* 195 USPQ 344 (1977).

125. See *Dick* v. *Lederle,* op. cit.

126. *In re Bergy* op. cit., p. 347.

127. Compare section 1 (2) (a) of the Patents Act 1977 (U.K.), which provides that a discovery is not an invention.

128. *In re Bergy,* op. cit. The court was split three to two in favor of patentability. As in the *Chakrabarty* case, the court accepted that the organism in question was new and unobvious and was not a product of nature as opposed to a "new manufacture" or "composition of matter" within the meaning of section 101.

129. The court also discussed a series of cases on the patentability of chemical compounds and noted that the legislative history of the Plant Patent Act 1930 was irrelevant as far as the *Bergy* application was concerned. See *In re Bergy*, op. cit., 351.

130. Ibid. The two dissenting judges simply restated the arguments that had been made by the board.

131. 283 US 1 (1930). The case concerned oranges that had been impregnated with borax.

132. 46 LW 4791 (1978). The Supreme Court decided this case five days before it heard the *Bergy* appeal.

133. Compare Section 1(2) (a) of the Patents Act 1977 (U.K.), which provides that a mathematical method is not an invention.

134. Section 101(1) of the Patents Act 1949 (U.K.) — the section under which the Chakrabarty application was granted in the United Kingdom — is not materially different from section 101 of the U.S. legislation. It should also be noted that if Congress wishes to prohibit the patenting of novel genetic organisms it will do so specifically as it did in the case of inventions used in the production of fissionable material. See the Atomic Energy Act 42 USC, section 2181. Presumably the policy objections that influenced the Supreme Court are at least partly based on the perceived administrative difficulties that would be involved in processing large numbers of patent applications in relation to organic inventions. This was certainly a possibility that disturbed the board in the *Bergy* case. However, that view is clearly not shared by the Department of Commerce. Early in 1977 the Commissioner of Patents and Trademarks issued the following statement: "In view of the exceptional importance of recombinant DNA and the desirability of prompt disclosure of developments in the field, the assistant secretary of commerce for science and technology has requested that the Patent and Trademark Office accord special status to patent applications involving recombinant DNA . . . including those that contribute to the safety of research in the field." See the *Federal Register*, January 13, 1977. Except in relation to safety technique applications, the order was temporarily suspended by a notice in the *Federal Register* of March 9, 1977. This action was taken pending a review of congressional criticism of selective processing practices. See "The Patenting of Recombinant DNA Research Inventions Developed Under DHEW Support," *NIH Recombinant DNA Technical Bulletin* 1, no. 2 (Winter 1978): 23.

135. Op. cit.

136. Compare Plant Varieties Act of 1973, op. cit.

137. See June Goodfield, *Playing God: Genetic Engineering and the Manipulation of Life* (New York: Random House, 1977), especially Chapter 9, and Clifford Grobstein, "The Recombinant DNA Debate," *Scientific American*, July 1977, pp. 29-30, cited in Delgado and Millen, "God, Gallileo and Government: Towards Constitutional Protection for Scientific Inquiry," *Washington Law Review* 53 (1978) p. 349. Also see Chapter 2 of this book.

138. "First Report of the Genetic Manipulation Advisory Group," op. cit., paragraph 9.3. Any person who sends a proposal to GMAG can request that it be classified as sensitive, but the final decision on this matter rests with the commissioner. Unfortunately, two GMAG members have refused to sign a declaration of confidentiality. See "Grabbing the Tiger," *New Scientist*, June 15, 1978, p. 730.

139. Compare the statements made by General Electric in Specification 1,436,573 (U.K.). op. cit.

140. For instance, Genentech, a company in California, has marketed genetically engineered somatostatin at a substantially reduced cost. See "US Company takes Genetic Engineering to Market," *New Scientist*, December 8, 1977, p. 619.

141. On the expiration of three years from the date of the sealing of a patent, the commissioner has a discretion to grant licenses to applicants who, for example, can demonstrate that a patent is not being commercially worked to the fullest extent that is reasonably practical in New Zealand. See Patents Act 1953, section 46. In the genetic engineering context, such licenses should be granted only after consultation with the National Advisory Committee as to

the suitability of the aspiring licensees. It is interesting to note that section 41 of the old legislation has been repealed by the Patents Act of 1977 (U.K.). Section 41 facilitated the granting of compulsory licenses in relation to pharmaceutical patents. A similar trend is evident in New Zealand. In June of 1978 the Pharmaceutical Manufacturers Association called for a four-year extension of the present 16-year patent protection period. The additional protection was sought only for pharmaceutical patents.

142. See sections 2 and 5 of the Accident Compensation Act 1972.

143. See the "Report of the Accident Compensation Commission for the Year Ended 31 March 1978," p. 20.

144. "Nuclear Power Generation in New Zealand," Report of the Royal Commission of Inquiry," April 1978, p. 226.

145. It has been suggested that unlikely accidents that will affect large numbers if they do occur are less acceptable to the public than accidents that are likely to cause little damage although they occur frequently. See ibid., p. 212.

146. The Earthquake and War Damage Act 1944, section 14.

147. See ibid., section 26(2) (bb).

148. (1869) L.R. 3 H.L. 330.

149. See the Paris Convention on Third Party Liability in the Field of Nuclear Energy and the Vienna Convention on Civil Liability for Nuclear Damage as discussed in "Nuclear Power Generation in New Zealand," op. cit., p. 221. The imposition of strict liability was proposed in clause 7 of the Recombinant DNA Research Bill, (S.621, 95th Cong., 1st sess., February 4, 1977), although it would not be backed by compulsory insurance. It is interesting to note that the Radiation Protection Acts 1965 do not deal with the question of compensation.

150. Spigelman, op. cit.

151. It can be argued that the existence of a mandatory insurance requirement eliminates any deterrent effect that the imposition of strict liability might have. However, as Fleming points out, there is no evidence that reliance on insurance cover fosters irresponsibility. See Fleming, op. cit., p. 11.

152. For an interesting discussion of criminal compensation, see "Restitution and Compensation," Law Reform Commission of Canada, Working Paper No. 5, 1974.

153. See clause 1797 of the California Biological Research Safety Commission Bill 1977.

154. Herbert L. Hart, "The Aims of the Criminal Law," in Law and Contemporary Problems, 40 (1958).

155. Compare The Radiation Protection Act 1965, section 26(1) (d).

156. See section 47 of the Clean Air Act 1972 and section 26(c) of the Radiation Protection Act 1965, which have been discussed in detail earlier in this chapter.

157. For example, Williams v. Huntly Borough [1974], NZLR 689.

158. The need for employee protection subsequent to the initiation of proceedings or assistance with enforcement has been considered in relation to emergency procedures and administrative remedies. Those comments are equally applicable in the criminal context. By way of analogy see section 150(1).(g) of the Industrial Relations Act 1973 and section 31(a) of the Human Rights Commission Act 1977.

159. Mahon, "Environmental Issues and the Judicial Process," New Zealand Law Journal (1976): 507.

160. Presumably the usual criminal defense standard of proof on a balance of probabilities applies. See R. v. Carr-Briant (1943) KB, 607.

161. See Spigelman, op. cit.

162. See section 26(2) of the Radiation Protection Act 1965 and section 56 of the Noxious Plants Act 1978.

163. Compare Radiation Protection Act 1965, section 27. In practice the project supervisor would be likely to enter into a contract of indemnity with his employer.

5

INTERNATIONAL CONTROL

In spite of the fact that all states are dependent on a shared environment, the international community has, in the past, been slow to recognize the need to regulate the harmful effects of new technologies. In most cases international action is preventive only in the sense that previous damage or disaster forms a basis for the promulgation of a code of conduct for the future. Regulation of the new technology of genetic engineering must not be allowed to conform to that pattern. The introduction of international measures that would reduce the risks that have given rise to concern in national settings is urgently required and should not be delayed until after a manifestation of perceived dangers.

To a certain degree, the nature and extent of the damage that could be caused by recombinant organisms cannot be predicted, but it is clear that, once released, self-replicating products of the technology of genetic engineering have the capacity to traverse national boundaries and to upset the ecological balance in distant states.

CUSTOMARY INTERNATIONAL LAW

If a genetically engineered virus or organism were to be released from one country into another, the recipient state could attempt to establish that the release constituted a breach of an obligation that was owed to it by virtue of a rule of international law — in this case an obligation not to release or allow the release of genetically engineered viruses or organisms into the territory of another state. Clearly, that particular obligation has not been recognized as a rule of customary international law. On the other hand it has been suggested that specific rights and their correlative duties exist at international law in the context of water pollution and air pollution. As far as the former is concerned, the decision of the tribunal in the *Lake Lanoux Arbitration*[1] has been used to provide support for the view that there is an international obligation not to discharge harmful agents into waterways that pass through other states. For example, O'Connell refers to the passage in which the tribunal noted that "it could have been argued that the works would bring about an ultimate pollution of the waters of the Carol or that the returned waters would have a chemical composition or a temperature or *some other characteristic* [emphasis added] which could injure Spanish interests."[2] He cites that extract as evidence of a rule of customary international law to the effect that a state that suffers damage as a result of the utilization of water by another state has a right to object,[3] and indeed the arbitrators imply that their decision might have been favorable to the Spanish litigant if such an argument had been raised. However, O'Connell omits to mention that the members of the tribunal stated that the argument that they had outlined was relevant because it would have given Spain a chance of success under the governing Additional Act of 1866.[4] As regards general principles, the tribunal could not find "in international common law, any rule that forbids one State, acting to safeguard its legitimate interests, to put itself in a situation that would in fact permit it, in violation of its international pledges, seriously to injure a neighbouring State."[5]

The *Nuclear Tests* cases also contain some interesting dicta in this context, although the International Court refused to deal with the merits of the Australian and New Zealand claims. For example, Judge Petren was of the opinion that there was no rule of customary international law whereby states were prohibited from conducting atmospheric nuclear tests that resulted in the deposit of radioactive fallout on the territory of other states. He purported to base that conclusion on state practice but yet he would not recognize the significance of certain relevant General Assembly resolutions except insofar as they "indicate the existence of a strong current of opinion in favour of proscribing atmospheric nuclear tests."[6]

On the other hand, in response to the Australian claim that there is a customary right to be free from the effects of atmospheric nuclear weapon tests conducted by other states, Judge Ignacio-Pinto (dissenting) stated that"each state . . . in the event of genuine damage or injury, owes reparation to the state having suffered that damage."* However, he was not prepared to

---

*It can be argued that the words *"genuine* damage or injury" are consistent with an intention not to preclude claims relating to intangible injury occasioned by a violation of sovereignty.

recognize an "existing legal means in the present state of the law which would authorise a State to come before the Court asking it to prohibit another State from carrying out, on its own territory, such activities, which involve risk to its neighbours."[7]

That supposed lack of legal means did not deter the tribunal in the *Trial Smelter* arbitration.[8] In that case it was decided that "under the principles of international law, as well as of the law of the United States, no State has the right to use or permit the use of its territory in such a manner as to cause injury by fumes in or to the territory of another or the properties or persons therein, when the case is of serious consequence and the injury is established by clear and convincing evidence."[9] The tribunal then went on to require the smelter to refrain from causing fume damage in the United States. For this purpose a system of control measures was prescribed although the smelter was not requested to cease operations altogether.[10] Aréchega is of the opinion that the *Trail Smelter* case is based on the *sic utere* maxim — a general principle of customary international law.[11]

He also suggests that the ordinary principles of state responsibility for unlawful acts form the basis of responsibility for the deposit of radioactive fallout on the territory of other states.[12] In the course of a discussion of abuse of rights he adopts the same approach in relation to "a state substantially affecting other states by emanations from within its borders . . . fumes, air or water pollution, diversion of waters . . . ."[13] Conversely, he cites the conduct of activities in outer space and the operation of nuclear power plants and nuclear ships as examples of situations in which the activities "causing or likely to cause damage are dangerous, but not unlawful."[14] He points out that in the absence of the treaties that govern those activities there would be no state responsibility for any damage that they might cause.

On that basis it is rather difficult to categorize genetic engineering activities causing or likely to cause damage in another state. They might be regarded as lawful* in the sense that it could be argued that they do not violate an obligation that is specifically established by a rule of customary international law.[15] On the other hand it would be possible to suggest that damage caused by genetic engineering is inevitably the result of air or water pollution. The fact that the pollution is caused by microorganisms or viruses rather than fumes should not bar a claim. In any case it is noticeable that Aréchega separates fume damage from air pollution.†

---

*If that were held to be the position under international law, the equitable doctrine of abuse of rights could apply to attach liability to a sponsoring state in the event that extraterritorial damage resulted from the activity.

†See Chapter 4, section on "Existing Law," for a discussion of domestic antipollution legislation and genetic engineering.

Even if it were established that harmful genetic engineering activities are governed by the rules of customary international law, it would still be necessary to ascertain whether a successful claim would depend on proof of fault as well as proof of damage. Jenks is of the opinion that state responsibility under customary international law has in the past been dependent on proof of fault.[16] He suggests that if there were a customary international law doctrine of liability for risk,* proof of fault would no longer be necessary in cases where ultrahazardous activities cause damage. Consequently, he claims that the *Trail Smelter* decision involves a clear application of the doctrine of liability for risk.[17] He states that when the doctrine applies, one of the bases of liability in international law is "shifted from fault (including negligence) to risk with a view to spreading more fairly the possible consequences of improbable but potentially disastrous activities . . . eliminating a burden of proof which, in view of the nature of the risk, the victim cannot reasonably be expected to discharge and, in many cases, could never discharge . . . ."[18]

Aréchega employs a less conventional line of reasoning. He argues that proof of fault is superfluous as far as state responsibility is concerned and that proof that an unlawful act has caused damage is sufficient unless fault has been explicitly recognized as an element of a particular rule of international law.[19] Accordingly he would not apply the doctrine of liability for risk to the *Trail Smelter* decision.

It is also possible to dispute the existence, as well as the effect, of a customary international law doctrine of liability for risk. Aréchega argues that the doctrine is not a general principle of customary international law.[20] On the contrary, he states that it applies only in circumstances that have been "previously and clearly defined by international agreement."[21] In support of this proposition he cites the Rome Convention on Damage Caused by Foreign Aircraft to Third Parties on the Surface,[22] the Paris Convention on Third Party Liability in the Field of Nuclear Energy,[23] the Brussels Convention on the Liability of Operators of Nuclear Ships,[24] and the UN General Assembly Resolution on Legal Principles Governing Activities of States in the Exploration and the Use of Outer Space.[25]

Jenks relies, inter alia, on the same treaties as Aréchega but he uses them to illustrate that the measure of express agreement on the existence of a doctrine of liability for risk in international law is such that strict liability for ultrahazardous activities seems to be evolving into a rule of customary international law. In addition, he notes that the term ultrahazardous activity "does not imply that there is a high degree of probability that the hazard

---

*The doctrine is also known, inter alia, as liability for the objective risks arising from ultrahazardous activities.

will materialize, but rather that the consequences in the exceptional and perhaps quite improbable event of the hazard materializing may be so far reaching that special rules concerning the liability for such consequences are necessary. . . ."[26] He is also prepared to recognize that "there may well be cases in which the current experiments of molecular biologists involve dangers which pose acutely the problem of liability for the objective risk involved in ultrahazardous activities."[27]

The development of the new technology of genetic engineering has meant that the problem to which Jenks referred in 1966 is now far more acute. On the other hand a customary international law doctrine of liability for risk does not yet appear to have crystallized.* Even more importantly, there is room for doubt regarding the application of recognized general principles of state responsibility to the release of genetically engineered viruses and organisms that traverse national boundaries. It is probable that the international tribunal that would be called upon to decide these matters in the event of an accident would share those uncertainties. Accordingly, a set of principles that govern state responsibility should be incorporated into the text of an international convention on genetic engineering.

POSSIBLE APPLICATIONS OF EXISTING TREATIES, CONVENTIONS, AND DECLARATIONS

The United Nations Conference on the Human Environment and the Declaration and Recommendations for Action that emanated from it[28] were among the earliest indications that the international community was willing to become actively involved in the protection of the environment. Prior to 1972 particular aspects of the environmental problem had already formed the focus of conventions, declarations, and multinational conferences,[29] but it was not until Stockholm that the importance of a broadly based international plan of action for the preservation of the environment was fully recognized.[30]

Part of the plan consisted of a declaration that there is a

common conviction that States have, in accordance with the Charter of the United Nations and the principles of international law, the sovereign right to exploit their own resources pursuant to their own environmental policies, and the responsibility to ensure that activities within their jurisdiction or control do not cause damage to the environment of other states or areas beyond the limits of national jurisdiction.[31]

---

*The international Law Commission is currently working on the clarification of the doctrine.

Accordingly, it was also declared that "states shall cooperate to develop further the international law regarding liability and compensation for the victims of pollution and other environmental damage caused by activities within the jurisdiction or control of such states to areas beyond their jurisdiction."[32] These words appear to extend to genetic engineering activities that cause damage in other states, although the recommendations for action contain references to toxic or dangerous substances rather than to harmful organisms.* In addition, principle 21 appears to constitute a recognition of general principles of international law that govern environmental damage. Conversely, principle 22 does not refer to general principles. Even the words "develop further" are unlikely to be construed as a reference to an existing rule of customary international law relating to liability and compensation for marine pollution and nuclear damage.†

The way in which the word disease is construed will determine whether certain existing treaties can be applied to genetic engineering. For instance, the International Covenant on Economic, Social and Cultural Rights‡ is concerned, inter alia, with the prevention and control of disease. Similarly, the International Agreement for the Creation at Paris of an International Office of Epizootics[33] relates to the establishment of an office that serves as a base for the coordination of research into contagious animal diseases.[34] The office, which is under the control of an international committee, is also responsible for informing governments of developments in the field of disease control.[35] A reciprocal obligation ensures that the governments involved inform the committee of the disease control measures they are implementing in their countries. In the absence of any definition or limitation of the words "animal disease" there is no reason why the provisions of the agreement should not be implemented if a genetic engineering accident produced or had the potential to produce harmful effects among animal populations.

The same considerations apply in relation to the International Plant Protection Convention,[36] which, unlike the International Health Regulations, is open-ended in the sense that its application is not limited to specific diseases.[37] The convention contains provisions concerning both national

---

*For example, see recommendation 71, which is to the effect that governments should use the best practicable means to minimize the release of toxic or dangerous substances into the environment.

†The need for protection of the environment is also recognized in the Declaration of Helsinki (adopted by the 18th World Medical Assembly in 1964 and revised by the 29th World Medical Assembly in 1975 in Tokyo). In the introduction to the declaration and in the context of research into human diseases it is stated that "special caution must be exercised in the conduct of research which may effect the environment."

‡Opened for signature in 1966 but not yet ratified.

control and international cooperation.* One of its primary objectives is the prevention of the spread of plant diseases across national boundaries.

The Budapest Treaty on the International Recognition of the Deposit of Microorganisms for the Purposes of Patent Procedure is also relevant to the technology of genetic engineering.[38] It is of particular interest in this context because it serves as an indication that, in the future, it might be possible to abandon disparate domestic patenting procedures in favor of a uniform international code relating to the criteria of patentability. Such a development would complete the trend that has been evidenced in the Patent Cooperation Treaty and the European Patent Convention and would lessen the likelihood of the rejection of a patent application in one country and its acceptance in another, as in Chakrabarty's case.†

On the other hand the treaty is not concerned with the types of microorganisms that can be patented and, accordingly, it cannot be claimed that the convention constitutes a recognition of the patentability of recombinant or novel organisms.‡ Nevertheless it is difficult, although not impossible, to rationalize the treaty with the restrictive U.S. views on the patentability of living things.[39]

Finally, the Convention on the Prohibition of the Development, Production and Stockpiling of Bacteriological (Biological) and Toxin Weapons and on their Destruction must be considered.[40] The convention, which has been accepted by 73 countries,[41] prohibits the development, production, acquisition, or retention of "microbial or other biological agents, or toxins *whatever their origin or method of production,* of types and in quantities that have no justification for prophylactic, protective or other peaceful purposes" (emphasis added).[42]

Those words almost certainly apply to genetically engineered organisms that are designed or retained for nonpeaceful purposes. However, the point is such a significant one that the issue of the applicability of the convention to recombinant DNA molecules has already been raised on more than one occasion in the United States.** For example, in 1975 Dr. David

---

*Each contracting party agrees to cooperate with the Food and Agriculture Organization in the establishment of a world reporting service on plant disease — see article 7.

†Ananda Chakrabarty's application in relation to a genetically engineered microorganism was granted in the United Kingdom but rejected in the United States. See Chapter 4, section on "freedom of information."

‡Compare article 53(b) of the Convention on the Grant of European Patents (European Patent Convention), which provides that European patents "shall not be granted in respect of plant or animal varieties or essentially biological processes for the production of plants or animals; this provision does not apply to microbiological processes or the products thereof."

**See Chapter 2, note 38, for a discussion of the irreversible damage that has been caused in the United Kingdom and the United States as a result of biological weapon tests that involved the use of naturally occurring microorganisms.

Baltimore, a pioneer in the field of recombinant DNA research, requested an opinion on this matter from James Malone, the General Counsel of the U.S. Arms Control and Disarmament Agency. Malone replied that the agency was of the opinion that the use of recombinant DNA molecules for other than peaceful purposes clearly falls within the scope of the provisions of the convention.[43] Article 10 of the convention is also significant although it does not relate to biological weapons. It provides for the fullest possible exchange of equipment, materials, and scientific and technological information concerning the peaceful use of biological agents and toxins.

The need for a uniform international regime for the purpose of controlling the production, storage, use, and spread of genetically engineered organisms is not met by the conventions and declarations that have been mentioned in the preceding discussion. Even if they were pieced together, the scattered provisions that may be applicable to the technology of genetic engineering would not constitute the complete and coherent system of worldwide control that is required by a technology that has such important implications for the international community.

SOME PROPOSALS FOR A CONVENTION ON GENETIC ENGINEERING

The Work of International Organizations

The inadequacies of existing treaties and the uncertainties that surround the application of customary principles reinforce the view that a comprehensive international convention on the subject of genetic engineering is urgently required. In fact members of the scientific community have already issued repeated calls for concerted international action in this area, particularly regarding the formulation of a set of guidelines that would govern the conduct of genetic engineering research.

Paul Berg was among the first to draw the attention of the public to the dangers of recombinant DNA experiments when he sought international agreement on a temporary moratorium on the research.[44] In pursuit of that goal he convened a meeting of the world's leading geneticists at Asilomar in February 1975. The conference can be regarded as an enormous success,[45] even if only in the sense that the publicity that surrounded it has led to a series of international conferences on the scientific aspects of genetic engineering,[46] and to the involvement of groups such as the International Council of Scientific Unions (ICSU), the European Molecular Biology Organization (EMBO), the European Science Foundation (ESF), the European Economic Community (EEC), and the World Health Organization (WHO). In addition, new interest groups have been formed especially for the purpose of securing international cooperation on the control of genetic engineering. The most notable of these is the Coalition for Responsible Genetic Research.

In accordance with its aim of providing technical advice to institutes that engage in genetic engineering research, EMBO has set up a purely scientific advisory committee that disseminates a considerable amount of technical information both to its members and to other interested organizations and individuals.* In addition, the committee has proposed that there should be a voluntary register of recombinant DNA research activities in Europe,[47] and that a single international genetic engineering "bank" should be established in order to lessen the risks that are involved in widespread cloning of the large quantities of genetic material that are required for certain types of experiments. That task could quite appropriately be carried out by the European Molecular Biology Laboratory (EMBL), which operates a large P4 containment facility in Heidelberg.† The committee has also expressed the opinion that both the British and the U.S. containment measures are satisfactory.[48] Nevertheless, it is inclined to favor the British regulations for the simple reason that its members are more likely to reach agreement on their implementation.‡

The European Science Foundation, which represents 18 countries, is one of the few organizations that has been prepared to become involved in an analysis of the nonscientific aspects of genetic engineering.** Its Working Group on Genetic Manipulation is composed of scientists, sociologists, and lawyers who, between them, have formulated recommendations for control. In essence, the foundation favors the adoption of a set of uniform regulations at the international level,[49] the compilation of national registers, the establishment of national committees with interpretive, advisory, and supervisory responsibilities, and the implementation of the necessary enforcement measures at the domestic level.[50]

The EEC has also become involved in discussions on the international control of genetic engineering — an involvement that was not unexpected, particularly in view of the fact that the Treaty of Rome provides that one of the reasons for the establishment of the EEC was to improve scientific

---

*The organization receives support from the governments of its member countries, which include 16 European states and Israel.

†The laboratory is funded by ten European governments and Israel.

‡EMBO is currently conducting genetic engineering risk-assessment experiments in conjunction with the Microbiological Research Establishment at Porton Down in England.

**Another notable exception is the Coalition for Responsible Genetic Research, which is headquartered in New York and derives some support from the Friends of the Earth organization. Nobel laureates George Wald and Sir MacFarlane Burnet are members of the coalition's executive committee.

cooperation.[51] The community's contribution has been made through the European Commission, which has ackowledged that it has a responsibility to ensure that the safety measures adopted by its member states are "harmonized" and that private industry adheres to the same standards as the public sector.[52] In order to discharge that responsibility, the commission has sought the views of the ESF and of special advisers who have been nominated by individual member countries. These advisers have been asked to give an opinion on the possible form and basis of a directive on genetic engineering.

A decision on whether to issue a directive has not yet been made, nor has the matter been placed before the Council of Ministers. Nevertheless, the special advisers have recommended that if a directive were to be issued it should be based on the definition of genetic manipulation that has been adopted in the United Kingdom (see Chapter 3, section on the United Kingdom) and that it should be accompanied· by a requirement that all researchers notify their proposed work to a nationally recognized committee.

The International Council of Scientific Unions is one of a small number of organizations that are involved in the review of the technology of genetic engineering on a worldwide rather than a regional basis.* Its involvement began in October 1976 with the establishment of the Committee on Genetic Experimentation (COGENE), which is comprised of 14 members representing eight nations, namely, the United States, the United Kingdom, the USSR, France, Japan, India, Switzerland, and West Germany. In addition observers from WHO, FAO, UNESCO, and UNEP attend its meetings.†

COGENE conducts risk assessment experiments, examines the large-scale industrial use of the technology, and sponsors lectures and training courses on the technology. By way of contrast with some of the other interested international organizations, it has been extremely systematic in obtaining the information on which its recommendations are based and, accordingly, it has made a useful contribution to the literature on the status of the technology and the diverse national attempts to regulate it. For example, the committee has distributed detailed questionnaires that by April 8, 1978, had been returned by 49 countries. The interim report, which was based on these statistics,[53] indicated that at the end of March 1978 at least 367 recombinant DNA projects were underway in 155 laboratories in 15

---

*This is also true of the Coalition for Responsible Genetic Research and the World Health Organization. In 1974 the International Association of Microbiological Societies set up an ad hoc committee on the risks involved in genetic engineering; however, it has not been particularly active.

†UNEP has not taken any other initiatives as far as genetic engineering is concerned.

countries.* Seven of these were at a high level of containment, 76 at a moderate level, and the remainder had been designated by the respondents as low-level experiments. Seventeen nations had drawn up guidelines for the conduct of the research, but it was significant that only five of these had prepared their own guidelines; the rest had adopted the NIH guidelines, either in toto or in modified form.†

The working party also noted the substantial national differences as far as supervisory procedures and methods of enforcement are concerned and concluded that "identical experiments require strikingly different levels of precaution in different nations. Moreover, although the general approach to biological and physical containment first proposed in the U.S. guidelines has been adopted by most other nations, the specific characteristics of each containment level differ markedly among different nations."[54]

In spite of the fact that the ICSU is in favor of the establishment of an international clone bank[55] and the introduction of uniform domestic controls, its valuable contribution in the field of dissemination of information has not been matched by the formulation of a set of guidelines that could be submitted for international agreement. In fact it cannot be denied that the work of the international organizations that have concerned themselves with the technology of genetic engineering has been rather unimpressive in the sense that detailed proposals for the regulation of the technology have not been forthcoming. As far as containment procedures are concerned, that general pattern has not been broken by the intervention of the World Health Organization.[56] WHO can, however, make a different kind of contribution and its involvement has already tended to shift the emphasis in the international genetic engineering debate from agreement by the community of scientists to agreement by the community of states.

From the outset, WHO has favored the continuation of genetic engineering research. In 1975 its Advisory Committee on Medical Research noted that "the continuation, under appropriate safeguards, of microbiological research, including genetic manipulation and cell fusion studies, is of the utmost importance for progress in medicine and public health."[57] Nevertheless, WHO has also acknowledged that it has "a clear duty to act as a worldwide coordinator and promoter of international collaboration in this field."[58] It has at least partially fulfilled that duty by preparing a series of reports that relate to the regulation of genetic engineering research.[59] The reports stress the need for the implementation of international controls[60] and the formulation of international contingency plans that could be set in motion in the event of laboratory or transport-associated accidents.

---

*This figure does not include protoplast fusion experiments.
†Thirteen countries expressed an intention to formulate guidelines in the near future.

In spite of its record in this field, or perhaps because of it (in the sense that WHO has consistently stressed the importance of the continuation of the research), WHO has been subjected to severe criticism by the Coalition for Responsible Genetic Research. The complaint arose out of the choice of delegates for an international conference on genetic engineering that was convened by WHO in Milan. The members of the coalition stated:

> We consider it deplorable that a three-day conference on recombinant DNA activities, sponsored by the World Health Organisation which is supported by public monies, has a partisan format of proponents that excludes other points of view held within the scientific and nonscientific communities. Nor does this March 29-31, 1978 conference provide for representatives of the public interest in one of the most controversial scientific and social issues of our times.[61]

Although the coalition's claims proved to be well-founded,[62] it must be remembered that, as its title suggested, the conference was intended to consist of a specialist discussion of scientific developments and their practical applications. If a conference were convened for the purpose of discussing the introduction of a much-needed set of international regulations, its format would presumably be quite different.

It has already been suggested that international regulation of the technology of genetic engineering can best be achieved by means of a convention sponsored by the World Health Organization.[63] If such a convention were to be adopted, the diverse national standards that have caused so much concern could be abandoned in favor of a uniform code. Izenstark elaborates on the advantages of WHO sponsorship. She notes that the organization could "exploit its prestige, strength and expertise to foster strict adherence to international regulations in genetic research." However, she does not fully justify her claim that a convention is the answer to the problems that are involved in the control of genetic engineering activities.

One difficulty that has frequently been alluded to in wider contexts is that "environmental problems characteristically require expeditious and flexible solutions, subject to current up-dating and amendments to meet rapidly changing situations and scientific-technological progress. In contrast, the classical procedures of multilateral treaty making, treaty acceptance and treaty amendment are notoriously slow and cumbersome."[64]

Although it requires some comment, that criticism is not difficult to answer. Shubber[65] deals with the point by citing the Convention for the Prevention of Pollution of the Sea by Oil[66] and the Brussels Convention Relating to Intervention on the High Seas in Cases of Oil Pollution Casualties.[67] He uses these conventions as illustrations of successful pollution control treaties. He also notes that it is possible to keep pace with scientific and technical developments by incorporating specific treaty provisions relating to revision procedures. Thus, for example, a specially appointed committee or organization could be empowered to review genetic engineering regulations at regular intervals and to make any technical changes that are required.

Similarly, it is clear that WHO involvement should be limited to sponsorship, technical assistance, advice, and recommendation. In the course of a general discussion of pollution control, Shubber refers to the possibility of promulgating international regulations under the authority of article 21 of the WHO Constitution. That article permits WHO to make regulations concerning sanitary and quarantine requirements and to implement other procedures designed to prevent the international spread of disease.[68] However, as Shubber notes, article 21 does not mention pollution control and accordingly WHO's authority to promulgate such standards is necessarily based on the implied powers theory and a liberal interpretation of article 21.[69]

That authority is clearly too weak for the promulgation of binding international genetic engineering standards and in this context it must also be remembered that provisions relating to the domestic implementation of a set of regulations relating to prohibitions, containment procedures,* emergency measures, and transportation[70] would constitute only one facet of a genetic engineering convention.†

The Likelihood of Agreement and the Terms of Agreement

There is a substantial likelihood that if a suitable international conference were convened the states that are engaged in genetic engineering research would reach unanimous or near unanimous agreement on the domestic implementation of a set of international regulations on genetic engineering. A majority of states appear to follow the NIH containment procedures. Even the European Molecular Biology Association, which has recommended the implementation of the British regulations, has acknowledged that they are not inherently superior to their U.S. counterparts. In addition, it should be possible to accommodate dissenting views on points of technical detail.[71]

While it may be relatively easy to secure agreement on a set of safety regulations,‡ it can be predicted that it would be somewhat more difficult to

---

*The prescribed procedures could be regarded as minimum requirements. It would also be possible to specify that all experiments that require P4 containment must be conducted in a single international facility. Similar suggestions have been made by EMBO, the ICSU, and the New Zealand Medical Research Council.

†It would be preferable to deal with issues concerning the patentability of genetically engineered organisms and viruses in a section of a separate patent treaty.

‡The regulations would be appended to the proposed convention on genetic engineering.

achieve a consensus on the delicate issues of liability and enforcement at the international level. As far as the latter is concerned, it is clear that it will be necessary to establish a mechanism that will enable a specially constituted international body to enforce the regulations by ordering a cessation of certain prohibited activities and directing the implementation of prescribed emergency measures in the event of an accident.

Izenstark proposed that recognized international inquiry procedures should be utilized for this purpose.[72] Such procedures would enable a signatory to the convention to complain to an established board of inquiry on genetic engineering if it were of the opinion that its interests were being harmed or were about to be harmed by genetic engineering activities in another state. The board could then propose remedial measures that might be acceptable to the parties. In the event of an emergency or a failure to reach agreement, the board would have the power to require the government of the defaulting signatory to enjoin the relevant prohibited activities.

Izenstark expresses the view that such a board should also be entrusted with the task of disseminating information on genetic engineering research and studying new developments in the technology. However, it can also be argued that that additional function would detract from the board's authority as a judicial body. In fact, if Izenstark's proposals were implemented, an unusual mixture of judicial, investigative, advisory, and regulatory functions would be vested in one body comprised only of scientists and laymen.

It would seem preferable to ensure that the enforcement agency is at least partially comprised of members who have some legal or arbitral experience. Purely technical guidance could be supplied by a specially constituted scientific advisory committee that could be elected by the signatories or appointed by WHO on the advice of the ICSU. That committee would be better suited to the tasks of reviewing regulations, sponsoring safety training courses, and distributing information on new scientific developments.[73]

Treaty provisions concerning state responsibility for environmental damage are not unprecedented, and although the technology of genetic engineering involves some unique risks, close analogies can be drawn between liability and compensation for air and water pollution[74] and responsibility for damage that is caused by genetically engineered organisms.

It is relatively clear that liability for extraterritorial damage that is caused by genetic engineering activities should be assessed on the basis of risk rather than fault. In other words, if a party to the convention could prove that genetic engineering activities that were attributable to another state caused it to suffer damage,* then that state should be able to recover

---

*As, for example, when an accident occurs during the transportation of recombinant material.

its losses without proving fault or a breach of the regulations.[75] Nevertheless, it will not often be easy for the claimant state to prove that the loss suffered was in fact caused by a genetically engineered virus or organism. The doctrine of *res ipsa loquitur* is not a recognized principle of customary international law and it would be unwise to attempt to shift the onus of proof by means of a provision in the proposed convention. Such a change would be oppressive,[76] and states would be unlikely to agree to its introduction. In the absence of that doctrine, proof that the damage complained of is attributable to particular genetic engineering activities would have to be established in much the same way that the United States proved that fumes released from the Trail Smelter in Canada were the cause of damage to trees and crops in the State of Washington.[77]

The problems that are involved in assessing damage[78] and insuring that the cause of damage has been correctly identified highlight the importance of the selection of an appropriate judicial body. An arbitral tribunal is the obvious choice and there would appear to be no reason why these questions could not be decided by the same tribunal as the one that has been proposed by the writer in the context of the enforcement of a set of international genetic engineering regulations.

Izenstark is of the opinion that two separate bodies should be established. The first would have an essentially scientific membership and would enforce the regulations. The second would take the form of an arbitral tribunal that would decide questions of liability and compensation. However, it can be argued that both functions are basically judicial and that both would require the evaluation of technical expert evidence.* Accordingly, it might be more convenient and less cumbersome to vest those functions in a single tribunal. This proposal also offers the likely advantage of lending a more authoritative tone to decisions that relate to the enforcement of the regulations.†

Finally, it is necessary to examine the best possible means of ensuring that a suitable source of compensation is available in the event of an accident. In this context it must be remembered that, at the national level, signatories to the proposed convention would implement the provisions of a set of international regulations. Those regulations would contain provisions relating to liability and compensation under domestic law. As far as that type of liability is concerned, the writer has proposed a system of operator's

---

*The special advisory committee that has been suggested could assist the tribunal with matters of scientific detail.

†Note that in the *Trail Smelter* arbitration (decision no. 2) the tribunal ordered the smelter to refrain from causing fume damage in the future. For that purpose it required the smelter to adopt additional emission control measures.

liability backed by a compulsory $5 million insurance. However, the convention should specify that, if domestic damage either does not occur at all or if it occurs and amounts to less than $5 million, the operator of an institution that is involved in genetic engineering activities should be liable to pay for the damage that those activities cause in other states, providing that his total liability does not exceed the $5 million maximum.

The proposed convention should also establish an international fund to help compensate those who suffer losses that are caused by genetic engineering activities.* The fund could be modeled on the relevant provisions of the Brussels Convention on Third Party Liability in the Field of Nuclear Energy (1963).[79] The parties to that convention have agreed that compensation of up to $120 million will be paid in the event of nuclear accident damage that is caused by a contracting party and that occurs on or over the territory or property of a contracting party or to the nationals of a contracting party on or over the high seas. The first $5 million will be provided by a domestic insurance scheme of the kind that is described above. The next $65 million are to be provided out of public funds supplied by the party on whose territory the relevant nuclear installation is situated. The remaining $50 million would be made available out of public funds contributed by the parties according to a formula that is based on the gross national product of the contracting states and the thermal power of their nuclear reactors.[80] If provisions relating to genetic engineering projects were substituted for references to thermal power, that type of formula could be used for the calculation of contributions to a fund that could be established under the proposed international convention on genetic engineering.

NOTES

1. *France* v. *Spain* [1957] ILR, 101.
2. Ibid., p. 123.
3. Daniel P. O'Connell, *International Law*, Vol. 1 (London: Stevens, 1970), p. 592.
4. *Lake Lanoux Arbitration*, op. cit., p. 123.
5. Ibid., p. 126.
6. The *Nuclear Tests Case (New Zealand* v. *France)* [1973] ICJ Rep. 135 and "French Nuclear Testing in the Pacific," Vol. 2, Ministry of Foreign Affairs, pp. 193–94.
7. *Australia* v. *France* [1973] ICJ Rep. 99, 131.
8. *U.S.* v. *Canada* (1939) 33 AJIL, 182 and (1941) 35 AJIL 684.
9. The *Trail Smelter* Arbitration (decision no. 2), op. cit., p. 716.
10. Although there was no decision of an international tribunal that authorized such restrictions, the arbitrators noted that as regards air and water pollution there were decisions of the Supreme Court of the United States that could be used as a guide in view of the fact that there was

---

*Compare the International Convention on the Establishment of an International Fund for Oil Pollution Damage (1971), which came into force on October 16, 1978.

no prevailing rule of international law to the contrary. See ibid, pp 714-16. It would seem that the emphasis that the tribunal placed on material damage in this case would, in future fume pollution cases, prevent the possibility of a successful claim in relation to intangible injuries caused by violations of sovereignty. An award of $78,000 was granted for damage to trees and crops; however, indemnity was not claimed for alleged harm to the health of American citizens. In order to satisfy itself that the sulfur dioxide fumes had caused the relevant damage, the tribunal consulted technical experts and considered prevailing wind conditions, air currents, surface temperatures, and other meteorological factors.

11. Eduardo J. Aréchega, "International Responsibility," in *Manual of Public International Law*, ed. Max Sorensen (New York: St. Martin's Press, 1968), p. 540. The decision is remarkably ambiguous, particularly as regards the application of domestic and international principles. See Alfred P. Rubin, "Pollution by Analogy: The Trail Smelter Arbitration," *Oregon Law Review*, 50 (1971): 259. For instance, the tribunal found that "neither as a separable item of damage nor as an incident to other damages should any award be made for that which the United States terms 'violation of sovereignty'" (see the *Trail Smelter* Arbitration, decision no. 1, op. cit., p. 208). That finding represents a heavy reliance on the arbitral compromise — a reliance that was not evident in relation to the tribunal's statements concerning Canadian liability.

12. Compare the opinion of Judge Petren in the *Nuclear Tests Case (New Zealand v. France)*, op. cit.

13. Aréchega, op. cit., p. 540.

14. Ibid., p. 539.

15. However, compare the United Nations Declaration on the Human Environment, UN Doc. A/Conf. 48/14 Rev. 1 (E. 73. II A.14, 1973), principle 21.

16. See also Roberto Ago, "Le Délit International," *Recueil Des Cours* 68 (1939): 419, 477.

17. See Wilfred Jenks, "Liability for Ultra-Hazardous Activities in International Law," *Recueil Des Cours* 117 (1966): 99, 122.

18. Ibid., p. 107.

19. Aréchega, op. cit., pp. 534-35.

20. See also Denis Lévy, "La Responsibilité Pour Risque en Droit International Public," *Révue Générale De Droit International Public* 65 (1961): 744, 758.

21. Aréchega, op. cit., p. 539.

22. 310, UNTS, 182-228.

23. (1961) 55 AJIL, 1082.

24. (1963) 57 AJIL, 268.

25. General Assembly Resolution XVIII, 1962.

26. Jenks, op. cit., p. 107.

27. Ibid., p. 169.

28. Op. cit. Held in Stockholm, June 5-16, 1972.

29. For example, see the International Convention for the Prevention of Pollution of the Sea by Oil (1956), 327 UBTS, 4 and the Paris Convention on Third Party Liability in the Field of Nuclear Energy (1961), 55 AJIL 1082.

30. See Ved P. Nanda, "International Environmental Law — A New Approach," *Journal of International Studies* 4 (1975): 101.

31. United Nations Conference on the Human Environment, op. cit., principle 21.

32. Ibid., principle 22.

33. Done at Paris on January 25, 1924. There are now 77 parties to the agreement. See 26 UST, 1840.

34. See ibid., article 4.

35. For this purpose the office publishes a monthly bulletin that contains details of the laws and regulations that are promulgated in various countries for the purpose of controlling contagious animal diseases, information on advances in epidemiological science, worldwide

statistics concerning the spread of disease among domestic animals, and bibliographical notices. See ibid., article 10.

36. 150 UNTS 67. Done at Rome on December 6, 1951. Seventy five states are parties to the convention.

37. See 21 UST, 3003, article 1.

38. Done at Budapest April 14–28, 1977, WIPO Doc. BP/PCD/1, May 31, 1977, reproduced in (1977) 16 ILM 285. As of May 23, 1978, the treaty had been signed by 18 states, although it is not yet in force.

39. See *In re Chakrabarty* 197 USPQ 73 (1978), and *In re Bergy, Coats and Malik* 195 USPQ 344 (1977).

40. 26 U.S.T., 583. Done in Washington, London, and Moscow on April 10, 1972. (Hereinafter referred to as the Biological Weapons Convention.)

41. France and China are notable exceptions, although in 1972 France passed domestic legislation that prohibited the development of biological weapons. See *Disarmament and Arms Control,* a Green Paper Prepared in Advance of the United Nations Special Session on Disarmament, New Zealand Ministry of Foreign Affairs, April 1978, p. 29. In addition 109 states are parties to the Protocol for the Prohibition of the Use in War of Asphyxiating, Poisonous or other Gases, and of Bacteriological Methods of Warfare, June 17, 1925.

42. See ibid., article 1(1). Article 1(2) relates to weapons, equipment or means of delivery designed to facilitate the use of such agents or toxins in armed conflict or for hostile purposes.

43. *National Institutes of Health Environmental Impact Statement,* on *NIH Guidelines for Research Involving Recombinant DNA Molecules,* October 1977, p. 39. That view was subsequently supported by Ambassador Joseph Martin in an address to the Conference of the Committee on Disarmament on August 17, 1976, ibid., p. 38.

44. Berg's plea was issued while scientists were preparing for an international conference on genetic engineering that was eventually held in Davos, Switzerland, on October 10, 1974. See *Genetic Engineering: Its Applications and Limitations,* proceedings of the Symposium organized by the Swiss Society for Cell and Molecular Biology and the Gottlieb Duttweiler Institute for Economic and Social Studies, 1974. The first public warning was issued by Dieter Soll and Maxine Singer in "Guidelines for DNA Hybrid Molecules," *Science,* 181 (1973): 1114.

45. The proceedings of the conference are reported in Paul Berg et al., "Summary Statement of the Asilomar Conference on Recombinant DNA Molecules," *Proceedings of the National Academy of Sciences,* 72 (1975): 1981.

46. See, for example, the *Report of the Falmouth Workshop on Studies for the Assessment of Potential Risks Associated with Recombinant DNA Experimentation, Recombinant DNA Technical Bulletin* 1, no. 1 (1977):19; the *Report of the International Symposium on Genetic Engineering Scientific Developments and Practical Applications, Recombinant DNA Technical Bulletin* 1, no. 3 (1978): 19; and the *Report of the U.S.-EMBO Workshop to Assess Risks for Recombinant DNA Experiments Involving the Genomes of Animal, Plant and Insect Viruses,* ibid., p. 24.

47. *National Institutes of Health Guidelines for Research Involving Recombinant DNA Molecules, Federal Register* 41 (July 7, 1976): 27906.

48. See *Nature,* 263 (1976): 719.

49. Like EMBO, ESF prefers the British containment procedures to those that have been adopted by the U.S. National Institutes of Health. Neither group has proposed a combination of the two systems. Compare the Recommendations of the New Zealand Medical Research Council Advisory Committee on Genetic Manipulation, 1977 (unpublished), p. 2.

50. See the *Recommendations of the European Science Foundation's Ad Hoc Committee on Genetic Manipulation* (Adopted by the ESF Assembly on October 26, 1976).

51. Treaty Establishing the European Economic Community (Treaty of Rome), 298 UNTS 3, 3.

52. See *Recombinant DNA Technical Bulletin* 1, no. 1 (1977): 32. Presumably the "harmony" requirement means that the relevant domestic standards need not necessarily be identical.

53. The ICSU Working Group on Recombinant DNA Molecules, "Interim Report to COGENE," June 1978.

54. Ibid., p. 3.

55. The ICSU places greater emphasis than EMBO on the dangers that are involved in the cloning and storage of large quantities of mammalian DNA. EMBO is more concerned with the need for a bank that would distribute bacteria, bacteriophages, and plasmids. See Colin Norman, "USA," Nature 263 (October 21, 1976): 630.

56. In these matters WHO prefers to rely on the advice of organizations like the ICSU. Compare WHO's recommendations on the transportation of materials used in recombinant DNA procedures, as described in "Facilitation and Safety in the International Transfer of Research Materials," WHO Report, CDS/SMM/76/1, Rev. 1., 1976.

57. Science 196 (1977): 4286.

58. "Towards More Effective Biomedical Research" WHO Chronicle, 30 (1976): 377.

59. Advisory Committee on Medical Research, "Report to the Director-General," June 13–17, 1977, ACMR 19/77, p. 14; "WHO Special Programme on Safety Measures in Microbiology — Progress Report 1977-78," CDS/SMM/78.4; and V. Sgaramella, "Public Health Aspects and Safety Regulations in Genetic Experimentation," CDS/SMM/78/5, 1978.

60. See CDS/SMM/78.4, op. cit., p. 2.2.10, where it is suggested that a coordinating group should be sent to the USSR to establish international guidelines for hygiene and the medical surveillance of laboratory personnel.

61. Francine Simring, "A Public Letter to the World Health Organisation," March 6, 1978.

62. See the "Report of the International Symposium on Genetic Engineering, Scientific Developments and Practical Applications," op. cit.

63. See Susan R. Izenstark, "Genetic Manipulation: Research Regulation and Legal Liability Under International Law," California Western International Law Journal 7 (1977): 203, 214.

64. See Paolo Contini and Peter H. Sand, "Methods to Expedite Environment Protection: International Eco-standards," American Journal of International Law 66 (1972): 38.

65. Sami Shubber, "The Role of WHO in Environmental Pollution Control," Earth Law Journal, 2 (1976): 363, 371.

66. Op. cit.

67. (1970) 9 ILM, 25.

68. See Shubber, op. cit., p. 372. The regulations bind members of the organization unless they give notice of a rejection or a reservation.

69. See also article 31(1) of the Vienna Convention on the Law of Treaties, in American Journal of International Law 63 (1969): 875.

70. The WHO Report on "Facilitation and Safety in the International Transfer of Research Materials," op. cit., would provide a very useful starting point for the transportation section of the proposed regulations, although part I.1 excludes plant pathogens from consideration. Part II 6 (e) is one of the most important recommendations. It consists of a proposal that "both the national and international transfer of materials of new genetic combinations should be in the form of nucleic acid molecules without live vector or host organisms wherever this is possible." See also the "Report of the WHO Special Programme on Safety Measures in Microbiology," op. cit., Annex 1.

71. The Soviet representative on the ICSU has stressed the need for the establishment of a global committee that could assist in the formulation of a set of international standards. See the ICSU "Interim Report to Cogene," op. cit., 12.

72. Izenstark, op. cit., p. 218. She cites the establishment of UN Military Observer Groups in India and Pakistan and the UN inquiry mechanisms that were used in the Middle East, and in South Africa in the context of violations of human rights.

73. It could also be given the responsibility of convening biennial international conferences for the purpose of reviewing the operation of the convention. Compare the Biological Weapons Convention, article 12, op. cit.

74. See Michael Hardy, "International Control of Marine Pollution," in *International Organisation: Law in Movement*, Essays in Honour of John McMahon, ed. James E. Fawcett and Rosalyn Higgins (London: Oxford University Press, 1974), p. 103.

75. For an analogous example of a convention that imposes strict liability, see the Rome Convention on Damage Caused by Foreign Aircraft to Third Parties on the Surface (October 7, 1952), 310 UNTS, 182-228. The Paris Convention on Third Party Liability in the Field of Nuclear Energy, op. cit., provides a useful precedent as far as limitation periods are concerned. It bars claims that are made after a period of not less than two years has elapsed from the date on which the person suffering nuclear damage had, or should have had, knowledge of the damage and its source.

76. In the *Corfu Channel* case the International Court was not prepared to acknowledge the principle in a situation where the defendant state did not allow access to evidence that might have established the cause of the relevant damage. However, the court did note that in such cases "the victim should be allowed a more liberal recourse to inferences of fact and circumstantial evidence." See [1949] ICJ Rep. 18, and Aréchega, op. cit., p. 538.

77. See the *Trail Smelter* arbitration (decision no. 1) op. cit., 194-99. The tribunal evaluated information on soil acidity, weather conditions, and the intensity and frequency of fumigations and then compared that evidence with statistics on crop yield.

78. See Bin Cheng, "Liability for Spacecraft," *Current Legal Problems* 23 (1970): 217 for an interesting discussion of the law that should be applied by tribunals that are called upon to assess damages. He concluded that the most suitable bases of assessment are the general international law principles of justice and equity as opposed to the relevant legal principles of the state that has either suffered or inflicted the damage in question.

79. See (1963) 2 ILM, 685. The convention is supplementary to the Paris Convention of 1960.

80. Fifty percent of the contribution is to be determined on the basis of the ratio between the gross national product of each party at current prices and the total of all the parties at current prices as shown by the OECD statistics for the year preceding the year in which the nuclear incident occurs. The other 50 percent is to be determined on the basis of the ratio between the thermal powers of the reactors situated in the territory of each party and the total thermal power of the reactors situated in the territories of all of the parties. See ibid., articles 3 and 12, and Jenks, op. cit., p. 137.

6

CONCLUSION

The accidental death of pine seedlings at the DSIR Plant Physiology Division in New Zealand and the smallpox fatality in England serve to illustrate that the immense benefits that are offered by the technology of genetic engineering are linked with unique risks. It is both desirable and possible to control those risks in a manner that is consistent with the development and exploitation of the technology.

The need for control has already been recognized in the form of statutory regulations promulgated in the United Kingdom and in Yugoslavia. Unfortunately, the regulations are extremely limited in scope and effect. In that respect they resemble the guidelines that have been formulated by a majority of the states that either sponsor or sanction the conduct of genetic engineering activities within their borders. Those countries have discovered that nonmandatory guidelines, backed only by peer group pressure and the threat of a withdrawal of funds, have not always been sufficient to ensure that the technology is controlled. For example, the U. S. National Institutes of Health guidelines have been breached by government researchers. The NIH response to the first recorded case of a breach was so ineffective that many researchers are now prepared to admit that the guidelines are being ignored.

In New Zealand the Irvine Committee has noted the suggestion that "specific precautions must be taken to prevent the release into the environment of cells with novel, and therefore unknown, genetic potential." The main recommendation of the committee is important because it envisages that the technology should be controlled through the provisions of a new statute. Beyond that the report is disappointing. The details of control under the proposed legislation were not examined and the problems that are involved in the control of nongovernmental institutions were left unresolved.

The need for a new statute for the purpose of controlling genetic engineering in the public and private sectors is highlighted by the fact that existing legal structures do not provide satisfactory means of controlling the technology of genetic engineering. The preventive remedies that are embodied in certain existing statutes can be said to apply to genetic engineering only on the basis of strained interpretations. Furthermore, speculative common law compensatory remedies represent extremely uncertain means of recovering potentially massive losses. Except in the case of work-related accidents, the Accident Compensation Act 1972 (N.Z.) is also unable to supply a solution.

A comprehensive statute on the subject of genetic engineering would offer the control that is urgently required. It would provide for the establishment of a 12-member National Supervisory Committee on Genetic Engineering. This commission could be responsible for advising the government on the promulgation and revision of a set of regulations under the new legislation. These regulations would establish a mandatory code of practice for the conduct of genetic engineering activities. Certain categories of experimentation might be prohibited. The second major function of the commission would be to promote safety and to publish information on the implications of genetic engineering research and on new developments. For those purposes, it should be empowered to conduct public hearings and to sponsor safety training courses.

A six-member National Advisory Committee should also be established. Its primary function would be to administer the licensing provisions of the proposed statute. It would inspect premises, consider research proposals, and advise the director-general on the granting of the project license, which would constitute a prerequisite to public and private sector genetic engineering activitites.

As far as individual companies and institutes are concerned, a three-member Employer Safety Committee headed by a biological safety officer could be considered. This committee would review the activities that are proposed within the organization and would determine whether proposals should be submitted to the National Advisory Committee. Its other functions would include the establishment of health monitoring programs for employees.

Effective regulation of the technology of genetic engineering will also depend on the establishment of an inspectorate vested with wide-ranging powers to enter premises, remove samples, and conduct investigations into the adequacy of facilities. The inspectors would be required to prepare a written report that would be forwarded to the Employer Safety Committee within a reasonable time. This proposal should not necessitate the creation of a separate new inspectorate. Inspectors seconded from the Occupational Health and Toxicology Branch of the Health Department could discharge the relevant functions.

The powers of the inspectorate must be matched by provision for the implementation of emergency procedures. Such procedures will be necessary if an accident occurs or if a dangerous situation is discovered in the course of an inspection. The procedures would relate to containment and eradication and could be modeled on section 30 of the Animals Act 1967 and section 12 of the Plants Act 1970, which relate to the proclamation of animal and plant disease emergencies. In certain situations the minister of health would be empowered to seize, destroy, or take control of dangerous pathogens. In the event that he failed to fulfill his duties under the act, concerned individuals should be able to institute proceedings for a remedy in the nature of mandamus. Conversely, relief in the nature of prohibition could be sought by those who claim to be affected by the proposed actions of the minister. However, applications for review are of limited use in view of restrictive standing requirements. Accordingly, the act should specifically provide that proceedings for injunctive relief can be instituted by the minister of health, on the advice of the Supervisory Committee, or by a member of the public. This remedy would be available against companies, private individuals, or governmental research institutes. A standard by which to gauge continuing or proposed activities should also be written into the statute. For example, it could be provided that either the minister or any person who has reason to believe that the continuation of certain genetic engineering activities would constitute a significant danger to the public health, or would not be in the public interest, could seek an injunction for the purpose of halting the relevant experiments. The problem of the potentially prohibitive cost of such an action could be resolved by the adoption of a special provision relating to fees and costs. Thus, in the event that the court determined that an action served a useful public purpose, the party who instituted the proceedings would be entitled to recover the costs of the litigation, including a reasonable allowance for fees paid to expert witnesses.

If the public is to participate in the control of genetic engineering, whether by way of injunction or other legal or nonlegal means, it will require access to information on the relevant techniques and their application in this country. However, industrial and research interests must also be protected and it will therefore be necessary to prohibit the unjustified disclosure of information that is obtained in the course of enforcement of the act. Although the grant of a patent eliminates the necessity for secrecy, the crucial period, as far as disclosure is concerned, may precede the approval of patent applications. This factor tends to mitigate against the early publication of the details of projects that are the subject of genetic engineering licenses. On the other hand it would seem that the New Zealand Patent Office is not likely to approve pending patent applications that relate to recombinant techniques and microorganisms.

In addition to their potential role in enforcing the legislation, members of the public will wish to be compensated if they suffer losses that are caused by the technology of genetic engineering. Neither existing common law remedies nor the Accident Compensation Act would be able to satisfy that need in a vast majority of cases. Similarly, the establishment of a contribution-based national fund does not appear to constitute an appropriate solution. A more compelling alternative can be found in the imposition of strict liability. Such liability would be limited to a statutory maximum amount and would be coupled with a duty to insure accordingly. The liability would attach to the sponsoring institute rather than to the supervisor of the relevant project and amounts in excess of the statutory maximum would be recoverable at common law.

The provisions of the legislation would be strengthened by the imposition of criminal liability on defaulting employees and organizations. A list of offenses, which would be set out in the legislation, would include liability for the breach of any provision of the act or any regulation promulgated under the act. The liability would not be strict, although it could be imposed on a "half-way house" basis. Thus it would not be necessary for the crown to prove that the accused intended to commit an offense, but the accused would have a good defense if he could prove that he did not intend to breach the act or regulations and that he took all reasonable steps to ensure that his actions would not constitute an offense.

Penal provisions would lend the sanction that nonmandatory guidelines and informal controls lack. The penalties would include the revocation, suspension, or limitation of licenses and the imposition of fines. The fines would also operate in a preventive manner in the sense that a continuing breach of the act would give rise to separate violations for the purposes of the penalty provision. Maximum penalties would vary according to the identity of the accused and the nature of the responsibility.

Finally, it is clear that genetically engineered viruses and organisms cannot be contained by national borders. At the international level more suitable forms of constraint are required. Unfortunately, general principles of customary international law and existing conventions do not provide the coherent system of control that is necessary. In the absence of such control an international convention on genetic engineering would be desirable. States could agree on a uniform code of practice for the conduct of the technology. The code could take the form of a set of international regulations that the parties would adopt at the domestic level.

Issues concerning enforcement and state responsibility are somewhat more complex than those that surround the adoption of regulations. Accordingly, it would be necessary to establish a tribunal to which states could refer their grievances. In addition to its arbitral functions the tribunal would be involved in the enforcement of the convention and the implementation of emergency measures in the event of an accident. It would also be able to draw upon the services of a technical advisory body that would be responsible for studying new scientific developments and disseminating information on the technology.

If all other national and international remedies fail, it should be possible for those who suffer loss as a result of the technology to resort to an international fund that could be established under the proposed convention on genetic engineering.

GLOSSARY

Bacteriophage	Bacterial virus.
Chromosome	A collection of genes. A rod-shaped body in the nucleus of a cell.
Clone/Cloning	A group of cells, all of which are identical because all have descended from a single common ancestor. By way of analogy, cloning a DNA molecule implies obtaining many identical copies of an original molecule by replicating it within a cell clone.
Cytology	The study of cells.
DNA	Deoxyribonucleic acid.
Endonuclease	An enzyme that cuts DNA sequences.
Enzyme	A protein that promotes chemical change without being consumed in the reaction.
Enzymology	The study of enzymes.
Eschericia coli	A bacterium that is found in the intestines of animals, including man. K-12 is a strain of this bacterium that is commonly used in the laboratory.

Most of these definitions are based on those that appear in the *National Institutes of Health Environmental Impact Statement on NIH Guidelines for Research Involving Recombinant DNA Molecules*, October 1977.

Eukaryote	A major group of organisms, the cells of which contain a well-defined nucleus and a cytoplasm containing other subcellular organized elements (for example, plant and animal cells).
Gene	A sequence that carries the information for a particular function and that is contained in a chromosome.
Genome	The total complement of genes or chromosomes in a cell.
Host cell	In the genetic engineering context, a cell into which foreign DNA or RNA is inserted.
In vitro experiments	Experiments with living systems or extracts from them, performed "in glass" as opposed to in the naturally living organism.
In vivo experiments	Experiments with living systems.
Plasmid	An intracellular genetic element that is extrachromosomal and is maintained in a stable manner by the cell. It is capable of replicating independently of the chromosome.
Prokaryote	A major group of living organisms in which the cell nucleus and cytoplasm are not clearly delimited (for example, all bacteria).
Protoplast	The actively metabolizing part of a cell, as distinct from the cell wall.
Protoplast fusion	A technique that induces cells to merge by the use of procedures that by pass normal sexual processes.
Recombinant DNA molecule	A molecule that is composed of parts of two or more molecules that were physically separate prior to recombination.

Restriction endonuclease   Cuts a double-stranded DNA sequence at specific sites in a manner that produces overhanging single-stranded ends that are complementary.

Shotgun experiment   An experiment in which all the DNA fragments cleaved by a restriction endonuclease are inserted into a vector DNA, which is then inserted into a cell. This is in contrast with other recombinant DNA experiments in which only selected fragments of DNA are inserted into a vector DNA.

Somatic cell   A cell that is not a reproductive cell.

Vector   A carrier of a recombinant DNA molecule; usually a plasmid or a bacteriophage.

Virology   The study of viruses.

Virus   A submicroscopic agent consisting of a core of nucleic acid surrounded by protein and capable of replicating only in a living cell.

INDEX

technology, 9–12; controlling genetic engineering, 12–18

physical containment, 7, 16, 19, 25, 45, 50

Plato, 1

Popper, 9

Prebble, 59

protoplast fusion, 7, 45

pseudomonas bacterium, 13, 76, 101-102

shotgun experiments, 6, 15

Shubber, 135–136

smallpox outbreaks, 16, 44, 144

Smith, 5

Snow, 10

somatic cell genetic research, 7, 58

somatostatin: production of, 13

Spigelman, 69, 109, 113

standing: proposals concerning, 93-95

Stetler, 27

strict liability, 109–110, 112–114

Sutton, W., 10, 17

Sweden: position as regards control, 52

Switzerland: position as regards control, 52

Thomas, 33

trespass, 78–79

Union of Soviet Socialist Republics: genetic research in, 4, 17, 53–54

United Kingdom: Diseases of Animals Act 1950, 46; Factories Act 1961, 46; Health and Safety at Work Act 1974, 45, 46, 48, 49, 53; Health and Safety (Genetic Manipulation) Regulations 1978, 48–50; Medicines Act 1968, 46; Patents Act 1977, 102; Porton Down, 16

United States: Administrative Procedure Act 1946, 28–29; Asilomar Conference, 24, 54; California Biological Research Safety Commission Bill, 41; Cambridge City Council (Mass.), 31–32; Clean Air Act, 36; Commission on Genetic Research and Engineering Bill, 39, 43; Federal Advisory Committee Act, 29; Federal Water Pollution Control Act, 36; Fort Detrick, 16, 29–30; Freedom of Information Act 1966, 33, 42; Harvard University (research at), 31–34; Hazardous Materials Transportation Act, 36; National Environmental Policy Act 1969, 27–29, 39; National Institutes of Health Guidelines, 24–27; Occupational Safety and Health Act 1970, 36; Patents Act 1952, 102–105; Plant Patent Act 1930, 102; Public Health Service Act, 35, 42; Recombinant DNA Regulation Bill, 42; Recombinant DNA Research Bill (No. 1), 37, (No. 2) 41; Recombinant DNA Safety Assurance Bill, 43; Resource Conservation and Recovery Act, 36; Toxic Substances Control Act, 36; University of California (San Francisco), 33–34

Van Leeuwenhoek, 1

Wald, 13

Watson, 5, 44

Weismann, 3

Wilkins, 5

Williams, 45, 47

Wilson, E., 4

Wilson, W., 59

Wolstenholme, 106

Yugoslavia: position as regards control, 53, 144

ABOUT THE AUTHOR

YVONNE M. CRIPPS is a law lecturer at Victoria University of Wellington, New Zealand, and a barrister and solicitor of the Supreme Court of New Zealand.

She has published widely on legal issues that arise in the scientific context. Her versatility is reflected in the range of different periodicals to which she has contributed. These include the *Modern Law Review*, the *International and Comparative Law Quarterly*, the *American Journal of International Law*, and the *New Zealand Science Review*.

Dr. Cripps holds an LL.B. (Hons) degree with first class honors and an LL.M. degree with distinction from Victoria University.